Household Sustainability

Household Sustainability

Challenges and Dilemmas in Everyday Life

Chris Gibson
Carol Farbotko
Nicholas Gill
Lesley Head
Gordon Waitt
University of Wollongong, Australia

Edward Elgar
Cheltenham, UK • Northampton, MA, USA

Published by
Edward Elgar Publishing Limited
The Lypiatts
15 Lansdown Road
Cheltenham
Glos GL50 2JA
UK

Edward Elgar Publishing, Inc.
William Pratt House
9 Dewey Court
Northampton
Massachusetts 01060
USA

A catalogue record for this book
is available from the British Library

Library of Congress Control Number: 2012952656

This book is available electronically in the ElgarOnline.com Social and Political Science Subject Collection, E-ISBN 978 1 78100 621 4

ISBN 978 1 78100 620 7 (cased)

Typeset by Cambrian Typesetters, Camberley, Surrey
Printed by MPG PRINTGROUP, UK

Contents

Tables

Acknowledgements

This book stems from a team research project, and from team efforts extending well beyond its named authors. We owe a substantial debt of gratitude to Elyse Stanes for project management and technical assistance across several ethnographic and quantitative activities over five years; and to Chantel Carr for her long hours doing crucial editing, research assistance and logistical work behind the scenes in book production. Chantel also provided important critical and technical feedback on draft chapters. Without these two individuals the book would simply never have been finished. We also thank Lucy Farrier and Alex Tindale for their research assistance and Chris Brennan-Horley for incisive comments on an earlier draft of this book.

Various chapters of the book draw on research conducted by our post-doctoral fellows and research students working at the University of Wollongong under the umbrella of our household sustainability project. Chapter 2 benefits from household ethnographies conducted by Bryce Appleby for his thesis, 'Skippy the "Green" Kangaroo: Identifying Resistances to Eating Kangaroo in the Home in a Context of Climate Change', as well as research on wheat conducted by Lesley Head in collaboration with Doctors Jenny Atchison and Alison Gates. Chapter 3 benefits from ethnographies of clothing consumption conducted by Elyse Stanes for her thesis 'Is Green the New Black? Questions of Sustainability for the Fashion Industry'. Chapter 4 benefits from quantitative and qualitative research by Candice Moy for her thesis on water tanks. Chapter 7 benefits from household ethnographies conducted by Gordon Waitt's advanced human geography (EESC 307) class of 2011. Chapter 10 benefits from driving ethnographies conducted by Theresa Harada for her thesis 'Driving Cultures and Climate Change: Bodies, Space and Affluence'. Chapter 13 benefits from ethnographic work by Erin Borger, for her thesis 'Dynamics of Extended Family Households: A Cultural Economy of Sustainability' (co-supervised by Dr Natascha Klocker). Throughout the book our treatment of gender issues benefits from the ethnographic work conducted by Vanessa Organo for her thesis 'Sustainable households: an analysis of time and gender in a household context'.

Beyond our household sustainability project are other collaborators whose work has influenced parts of the book: Pat Muir for collaborative work with Lesley Head on backyard gardens; Peter Osman (CSIRO) for collaborative work with Nick Gill on solar hot water; and Paul Cooper from UOW's Sustainable Buildings Research Centre for collaborative work on retrofitting homes.

For leads, conversations and resources helping us think through the issues (and beyond specific projects listed above) we thank Jenny Atchison, Paul Cooper, Mark Davidson, Eliza de Vet, Kelly Dombroski, Rosie Emery, Alison Gates, Leah Gibbs, Charlie Gillon, Andrew Gorman-Murray, Russell Hitchings, Natascha Klocker, Ruth Lane, Beth Laurenson, Candice Moy, Pat Muir, Vanessa Organo, Emma Power, Kate Roggeveen, Justin Spinney, Emma Waight, and John Wright. We also thank our families for their patience and equanimity at home – and for daily debating of our own sustainability dilemmas.

This book stems from an Australian Research Council (ARC) Discovery Project 'Making less space for carbon: cultural research for climate change mitigation and adaptation' (DP0986041) and an ARC Future Fellowship to Chris Gibson: 'Crisis and change: cultural economic research on the adaptability and sustainability of Australian households' (FT0991193). Its writing was also facilitated by an ARC Laureate Fellowship to Lesley Head (FL0992397).

Finally, we would like to thank all the householders who were research participants in the various sub-projects, surveys and ethnographic investigations featured in this book. Without their generosity and willingness to allow us into their homes, gardens and garages (and into their water and electricity bills), there would be no stories for this book.

Abbreviations

ABS	Australian Bureau of Statistics
ACF	Australian Conservation Foundation
AMTA	Australian Mobile Telecommunications Association
ARC	Australian Research Council
CFCs	chlorofluorocarbons
CO_2	carbon dioxide
CO_2-e	carbon dioxide equivalent
CRT	cathode ray tube
CSIRO	Commonwealth Scientific and Industrial Research Organisation (Australia)
DCCEE	Department of Climate Change and Energy Efficiency (Australia)
DEWHA	Department of the Environment, Water, Heritage and the Arts (Australia)
DFT	Department for Transport (United Kingdom)
DITRDLG	Department of Infrastructure, Transport, Regional Development and Local Government (Australia)
DSEWPC	Department of Sustainability, Environment, Water, Population and Communities (Australia)
EIA	Energy Information Administration
EPA	Environment Protection Agency
FIRA	Furniture Industry Research Association
GGE	greenhouse gas emissions
gha	global hectare (area measurement in ecological footprint analysis)
GPS	global positioning system
GWP	global warming potential
HC	hydrocarbons
HDPE	high-density polyethylene (plastic)
HFCs	hydrofluorocarbons
HVAC	heating, ventilation and air conditioning
IATA	International Air Transport Association
ICAO	International Civil Aviation Organisation
ICT	information and communications technologies
IEA	International Energy Agency

ITF	International Transport Forum
ITU	International Telecommunications Union
kWh	kilowatt hours
LCA	lifecycle assessment
LCD	liquid crystal display
LDPE	low-density polyethylene (plastic)
MEA	Millennium Ecosystem Assessment
Mt	megatonne
OECD	Organisation for Economic Co-operation and Development
OICA	International Organisation of Motor Vehicle Manufacturers
OPEC	Organization of the Petroleum Exporting Countries
PBT	persistent bioaccumulative toxin
PET	polyethylene terephthalate (plastic)
PP	polypropylene
TWh	terawatt hours
UN DESA	United Nations Department of Economic and Social Affairs
UNEP	United Nations Environment Program
VCOs	voluntary carbon offsets
WHO	World Health Organization
Wi-Fi	wireless local area network
WMO	World Metereological Organization
WSSCC	World Supply and Sanitation Collaborative Council

Introduction

Is it easy being green? This book discusses sustainability dilemmas faced by households in the course of everyday life. Contrary to the often-espoused rhetoric that being green is 'easy', household sustainability is rife with contradiction and uncertainty. While gains can be made through some actions, they must be traded off against other losses – and sometimes whether these trade-offs are worth it remains unclear. Is it worse to waste the water to rinse out tin cans than to put them in the recycling bin dirty? Is it worse to use plastic supermarket bags for bin liners, or to take reusable green bags to the supermarket but then buy dedicated bin liners (Chapter 9)? Some of these are empirical questions dependent on the chemistry of plastic. But beyond the need for better calculations of these sorts of trade-offs, are dilemmas of everyday practice. How much time do well-intentioned people spend thinking these choices through, debating them within a household? Which members of households undertake this 'thinking work'? Who feels guilty? What could be the outcomes if the same amount of effort was invested elsewhere? Some purportedly 'sustainable' behaviours such as eating local food may in fact prove more damaging than eating imported food, because 'food miles' associated with transportation of fruit and vegetables can be only a minor component of total carbon impact (Chapter 2). Two seemingly identical tomatoes on a greengrocer's counter may involve diverging 'journeys' to the shop, with quite stark differences in carbon impacts depending on a range of factors. Myriad dilemmas and alternative considerations prevail over basic, everyday behaviours and actions.

This book stems from our own frustrations seeking to navigate such dilemmas – but also from a concern that a simplistic set of assumptions about what it means to 'be green' are becoming quickly entrenched. The risk is that simple prescriptions overlook the complexity of dilemmas that surround everyday household choices and behaviours. Our desire to challenge assumptions is not because we are sceptical about climate change – we are not. Nor are we unsure about the need to act urgently. Something has to change, in the face of our current reliance on torrents of resource extraction, production and consumption. The household might be one place to imagine such change. In this book we engage critically, and constructively, with this proposition.

Whether households can readily make a difference is at one level a question of lifestyle and material comforts. Industrial capitalism removed the masses from peasantry and furnished people (or at least, middle-class people) with material possessions and pleasures – even luxuries. People have been promised, and have come to expect, a level of everyday comfort and gratification. Yet this is relatively recent; as little as a hundred years ago people struggled to stay warm or cool, to find and keep fresh food, to access reliable safe water. Billions still do. Nevertheless, today more people than ever before enjoy televisions and international holidays and stockpile large wardrobes of clothes. Such accoutrements are specific to wealthy groups who constitute the minority of the world's population. At various points in the book we confront cultural norms of lifestyle and consumerism that are linked to troublesome environmental problems – and question whether they can be easily unsettled.

The book is unapologetically focused on affluent countries, and especially on 'Western' cultures (in Europe, North America and Australasia). This focus stems from the now accepted wisdom that affluence is intimately linked to escalating greenhouse gas emissions. Households in the global south face entirely different challenges for 'sustainability': for them, consumption may simply mean survival. Our choice to focus largely on the rich world is not intended to cast aside the global south as 'other', but rather to position affluent lifestyles squarely at the centre of the problems of climate change and sustainability. We are also aware that Western cultures and affluent worlds are not one and the same thing. With some exceptions (for example toilets, Chapter 6, and retirement, Chapter 18) we do not focus on Japan and South Korea, although they are affluent. We acknowledge the importance of better understanding household dynamics everywhere, and at times we do look to the global south for insights on how things could be alternatively done. But throughout, we make the point that it is households in those parts of the world already making the largest contributions to greenhouse gas emissions where the most urgent changes are needed.

We are also aware of the various definitions of 'sustainability' and the ideological purposes to which that term is put (Davidson 2010). We focus here on material environmental impacts: greenhouse gas emissions, water and energy use, biodiversity and ecosystem loss, soil erosion and pollution. We also link sustainability to related social justice concerns with equality and exploitation.

We are not for a moment suggesting going 'backwards' to a quaint, pre-industrial age without medicines, sanitation or large-scale infrastructures. Indeed, the need for more integrated, systemic approaches to food security, public health and well-being is heightened with climate change. Rather, our premise is that progress towards a future that ensures prosperity and quality of life will require more than technological or organizational changes. We somehow also need to transcend seemingly intransigent cultural norms. The household is an especially vivid scale at which such norms are manifest.

At another level, households face dilemmas of practice and circumstance: being green may not in fact be so easy, or its choices self-evident. While in some countries energy and water-use labels are present on new appliances, for the bulk of household consumables their carbon footprint or broader environmental imprint is unknown. Throughout the various chapters in this book we synthesize existing research on the carbon footprints and lifecycle impacts of products – to gain a more complete picture of their environmental imprint. Figuring out a product's true origin and impact amidst marketing spin is difficult enough for just one item, let alone for all the stuff that enters a typical house. People juggle competing priorities and pressures and imperfect information.

Ours is also a very geographical take on household sustainability – for we are all geographers working at the intersection of environmental science, cultural studies and political economy. Our perspective emphasizes how geographical context, socio-economic circumstances and cultural meanings vary. Air conditioning might be justifiable for homes in which infants, the sick or elderly need to stay cool to remain healthy (Farbotko and Waitt 2011), while in other places and circumstances be challenged as an energy-sucking feature of suburban life – an unsustainable norm or even addiction (de Vet 2012) (see Chapter 5). Household use of water and relationships with water-using appliances (such as toilets and solar hot water heaters) vary from dry to wet and cold to warm places, but also with extent of government regulation, age, class, and even religious belief (Chapters 4, 6 and 15). How we deal with the death of loved ones and dispose of their corpses is deeply linked to cultural norms that might not be easily transformed (Chapter 19). Drawing on global examples we seek to *place* the debate about household sustainability rather than speak in generic terms.

This book seeks to progress our collective response to climate change by airing and discussing such dilemmas of sustainability. Only if we properly grasp the dilemmas of sustainability can we hope to go some way to overcoming them.

STRUCTURE OF THE BOOK

The book tackles what we see as flashpoint dilemmas. Each chapter is in turn framed around an essential task, everyday item or practice of household life. In many cases, there are no easy answers as to which course of action is most appropriate. By bringing light to the dilemmas we aim to increase understanding of both the barriers to household sustainability and the 'unsung' sustainability work being done by householders.

Beyond this introductory chapter (in which we set the scene in relevant social science, environmental and policy literatures), we take each flashpoint example of everyday life in the household and frame a concise discussion around them. We draw on international research to grasp the most pressing sustainability problems surrounding flashpoint dilemmas. Issues of the design and manufacture of objects are discussed, alongside available research documenting known energy, water and other resource use implications. Each chapter then moves to discuss fresh insights gleaned from fine-grained cultural research – international where available as well as from our empirical research projects on everyday household sustainability. These latter projects include data from a large survey on everyday household practices, goods, consumption and attitudes in Australia, and in-depth ethnographic material from longitudinal research, where we interviewed households over a two-year period. These studies provide concentrated, fine-grained examples to illustrate more general discussion on the dilemmas of sustainability.

We could not cover every dilemma, or every household activity. There are things we know that have received insufficient attention but would require another whole volume of dilemmas: tampons, weddings, coffee, wine, going to school, medicine, pets, showers, renovations. One of the dilemmas (in the book as in everyday life) is which dilemmas matter more than others. Because each chapter is concise we have made choices to focus on especially vivid examples (such as nappies, in Chapter 1) or insightful vignettes from ethnographic research (as with water tanks, Chapter 4). Some chapters slice a particular 'angle' through an issue and therefore cannot cover others. Some by necessity have a more technical flavour (as with solar hot water heaters, Chapter 15) while others confront knotty moral conundrums (as with toilets, Chapter 6). Nevertheless, throughout we share a sensibility attuned to dilemmas and contradictions, frictions and pathways for traction.

THE HOUSEHOLD SCALE

A central premise of this book is that the household scale matters – but in complex ways – to the human response to climate change, and more broadly to problems of conservation, biodiversity protection and sustainability. Households make sense both to the people who live in them, and to government policy makers, as foundational social units (Lane and Gorman-Murray 2011), and are frequently used to measure the consumption of energy, water and materials. In affluent urban societies households are an increasing focus of government policy on sustainability. An expanding research literature considers the household as a crucial scale of social organization for pro-environmental behaviour (Reid et al. 2010; Gibson et al. 2010; Lane and Gorman-Murray 2011; Tudor et al. 2011).

Across the global north governments at all scales have, to varying degrees, aligned households with sustainability. Governments have funded support for solar panels, home insulation, water tanks, light globes and shower timers in efforts to reduce greenhouse gas emissions, water and energy consumption. Local programmes, typified by the Sustainable Illawarra Super Challenge in our home city, Wollongong, encourage householders to become more environmentally sustainable by refusing plastic bags, composting, establishing vegetable gardens and catching public transport. Marketing materials use phrases like 'take the challenge to see just how easy it is to take control of your ecological footprint. You'll be surprised at how little time it takes to make a difference ... and how good it makes you feel!' (Sustainable Illawarra 2008).

Despite the enthusiasm to contribute to sustainability goals, such policies do not always have the intended outcomes. Smart meters do not challenge practices that householders consider non-negotiable (Strengers 2011). Water tanks do not save as much water as predicted (Moy 2012) (Chapter 4). Education programmes emphasizing that 'it's easy being green' understate the amount of domestic labour involved, and sidestep the question of who does the work (Organo et al. 2012). Residential energy consumption continues to rise, due to a combination of bigger homes containing more appliances and computer equipment, a growing population and a declining number of people per household (Australian Bureau of Statistics 2009). Furthermore attitudes and practice often do not match. Some of the most avid water savers vehemently express anti-green attitudes (Sofoulis 2005, 447), drawing instead on a rhetoric and identity of frugality (Chapter 4). At various points in the book we see the sustainability work being done by low-income households who do not necessarily identify as green, but who do consume less.

It is a truism that sustainability challenges are complex, but in this book we contend that the conceptualization of the household in environmental policy has not been complex enough. Many policy approaches treat households as black boxes – freestanding, bounded social units operating only at the local, domestic scale. The difficulty of tracking the contribution of households to their nations' greenhouse gas emissions provides an illustration of this complexity. In Australia, calculations vary depending on the assumptions made about where responsibility is to be attributed: 13 per cent if only direct energy use within the household is considered, and 56 per cent if the emissions embedded in externally produced goods and services consumed in the household context are included (ABS 2003). As the growing literature on carbon and other ecological footprints makes clear, this variation is partly an issue of data measurement and scale (Wilson and Grant 2009). We argue here that there is also a broader conceptual challenge: how should we think about configurations of people and material things whose social and ecological relations are diverse, shifting and complex?

Our response to this challenge develops what in the social sciences and humanities is known as a *relational* approach. The basic tenet is that the household, as a social-geographical scale, is entangled relationally with other actors at 'larger' and 'smaller' geographical scales. Relationality challenges the idea that we can 'identify discrete scales from which causes originate and at which effects are felt' (McGuirk 1997, 482). Thinking of households as discrete entities forces 'processes, outcomes, and responses... into distinct "boxes"' (McGuirk 1997, 482). In contrast, relational thinking encourages analysis of a variety of actors, big and small, human and non-human, inter-twined in patterns of power relationships (Bennett 2010). Multiple materiali-ties and networks make up the family home (Kaika 2004; Blunt 2005; Head and Muir 2007a).

This sort of thinking alters where one might go looking for the root causes of problems, and how one might subsequently trace actions and responses. For instance, are gatekeepers – educators, newspaper editors, carbon offset retail-ers, energy-saving websites, solar hot water installers, electrical switches – as important as individual household actors? Gatekeepers mediate relationships and flows between scales and things. The relationships between scale and order, or scale and causation, should not be assumed but be the subject of empirical enquiry. Gille and O'Riain (2002, 286) make the further point that level of analysis should not be confused with the level of abstraction – the global is not necessarily universal, and the local is not necessarily particular (Hulme 2010). Actions at the 'global' scale do not necessarily supersede or cascade 'down' to those 'locally' (Bulkeley 2005). To say that scale is both socially produced and relational does not deny that particular scales can become fixed, reproduced, and influential. Rather, geographical scales are entry points into the jumbles of things, relationships and differential capacities for agency that make and remake the world (Waitt et al. 2012a).

The household is one such scale. It is a common sense unit that operates as a 'spatial fix' for a variety of policy and social logics, but one that *becomes*, rather than merely exists *a priori*. There is nothing 'natural', for instance, about gender roles within families – these come into being through social norms and practices, in turn shaping deeply who does the work of sustainabil-ity in the household (Organo et al. 2012; see Chapters 1, 2, 7, 12). We explore households as social assemblages with variable gender, age, class, ethnic and familial structures (Blunt and Dowling 2006). The family with children, the student shared household, the extended family or the retired couple will all experience and respond to climate change and sustainability concerns differ-ently, as will homeowners, private and public renters, and unit and house dwellers (Farbotko and Waitt 2011; Klocker et al. 2012). Households will also argue within themselves over the best courses of action, as discussed here in relation to nappies (Chapter 1), clothes wash temperatures (Chapter 3), how to keep warm in winter (Chapter 5) and what stuff to keep or to throw out (Chapter 13). Households are homes in which social relations are the core human concern; in which families bond, people invest emotions and undertake all kinds of identity work beyond the putatively 'environmental'. Furthermore, as Hinchliffe (1997) argued, homes are understood as a refuge or haven from the problems of the world, confounding its potential role as a site of climate change mitigation via changes in household practices.

What exactly constitutes a 'household' is increasingly contested, as demographic and physical definitions (based on family units and/or buildings) are supplanted by notions of the household as networks of connections that mediate relations with other publics, with nature, with outside institutions (Reid et al. 2010, 318). The black box is revealed to contain humans and diverse non-humans, its own cohabiting things, complex politics and practices. The challenge is, to quote Ruth Lane and Andrew Gorman-Murray (2011, 2), 'to consider the operations of the household in terms of interactions between different animate and inanimate entities'. Homes are containers for appliances, pets, stuff – none of which we should take for granted as inanimate or powerless, or as disconnected from wider movements and flows. The black box is porous. Home spaces and the people and things that live in them are inextricably linked into the social, technological and regulatory networks that make up suburbs, cities, regions and nations.

Granularity is therefore needed to disentangle the complex assemblage that is the household. This is one reason why we have chosen to organize the book into digestible vignettes around everyday things and concerns. At the same time, however, we are mindful of Diana Liverman's (2008) question about how to 'upscale' social research to match scientific datasets, if social and cultural researchers are to participate in policy debate over climate change. This book seeks to bring together a more complex overall picture of households and sustainability, assembled from synthetic discussion of the atomized things and practices that typify home life.

It is important not to assume that households in the affluent West are powerless. The issue is how power is exercised in relationships between actors. Some of these actors include the state, infrastructure providers and planners, while in other directions relationships exist with appliances (and their manufacturers), retailers, corporations, communities – with even water and energy itself. Households have capacities and can generate traction along diverse pathways – sometimes informal and unheralded. Some such pathways are gradual, opening possibilities for change where immediate overhaul is unlikely, such as with norms of toileting (Chapter 6) or Christmas (Chapter 17). Other pathways depend on capacities to adapt quickly – as with responses to financial hardship (such as finding ways to heat bodies in winter without need for electric heaters – Chapter 5). In certain circumstances resisting change can be productive – as when ignoring the imperative to consume clothes or furniture as fashion items (Chapters 3, 8) or not turning on the air conditioner (Chapter 5). In this understanding, the 'local' – which can include the household – does not just feed into pre-existing scales of something bigger in accumulative fashion, but rather can itself be a generative site of creative possibilities. We therefore put to the test this vision of transformation and change – in which the household might be one vital, if not straightforward, site of action.

PLACING THE HOUSEHOLD IN CLIMATE CHANGE AND SUSTAINABILITY DEBATES

Consuming Citizens

How have households been placed in debates about sustainability and climate change?

First, there has been a considerable focus on consumption, particularly in official and corporate programmes that have sought to change household behaviour on environmental grounds. For Slocum (2004), climate change programmes operate within the neoliberal state, assuming citizens are mere passive 'consumers'. Sustainability however relies on the notion of the 'responsible, carbon-calculating individual' (Dowling 2010, 492), constructed in climate change campaigns as the new ideal citizen-consumer (Rutland and Aylett 2008). Promotion campaigns risk treating people merely as consumers rather than citizens or active subjects negotiating everyday lives (Burgess et al. 2003, Malpass et al. 2007).

The assumption is that many households are 'doing the wrong thing' vis-à-vis sustainability. Households are, to use a somewhat unwieldy phrase from governmentality theory, *problematized* as sites of excessive consumption, urban sprawl, and overuse of energy and water – no more so than in the typecasting of new-build homes as 'McMansions' (Nasar et al. 2007; Dowling and Power 2011). Ideal sustainable citizens transform themselves in response to this problematization – weaning themselves off high energy lifestyles. Smuggled into this narrative is, however, a subtle shift that places the burden of responsibility for change onto householders, rather than onto governments or corporations.

Campaigns framed around everyday technologies such as recycling bins, energy-saving light bulbs and shower timers trigger discussions of environmental ethics within households. In turn they 'produce' sustainable citizens who buy the right things and install new green technologies around the home, reduce car dependency, lower heating thermostats or raise air conditioning settings (Hobson 2006). However as both Davidson (2010) and Knox-Hayes (2010) argue, this keeps intact the institutions of capitalist democracy – markets, corporations, governments – and emphasizes that households must govern their own actions in order to become 'proper' sustainable citizens.

How households react is, nevertheless, unpredictable. As Scerri's (2011, 175) survey work with households in Melbourne demonstrates, 'householders are rejecting official claims that "rational" sustainable consumption choices and self-regulatory approaches will achieve the kinds of changes that sustainable development necessitates'. Overall, the conclusion from research assessing education campaigns is that promoting public awareness of global risks is inadequate to change behaviour (Lorenzoni et al. 2007, Robbins 2007). Humans are social beings acting in the contexts of communities and landscapes, not atomized individuals driven solely by rational economic or environmental preferences (Hobson 2002; 2003). Household adaptations vary in different contexts (Potter and Starr 2006; Hulme 2008), and are more than just the amalgam of individuals' 'energy minimizing' actions (Gibson et al. 2010).

We do not assume that any particular actions and behaviours by householders are necessarily 'better' than others. Imperfect information about total environmental impact of commodities and actions prevents this (Gibson et al. 2011). Some of the dilemmas occur when 'sustainability' behaviours are at cross-purposes, but we also are conscious that green morality is closely linked to green consumerism (Chapter 3). Rather than promote this, we seek to situate households in complex scenarios where competing interests, trade-offs and moralities collide.

Environment is just one line of responsibility being juggled in acts of consumption, which necessarily serve different anticipated needs (Dowling 2000; Burgess et al. 2003). Urging households to behave responsibly has its limits, with what constitutes 'responsible' behaviour developed within social practice, rather than abstractly distant from it. In the absence of mandated action or new infrastructure that shifts what is 'normal' (Shove 2003), we discuss how household consumption can remain unchanged for reasons that are perfectly ethical from that household's point of view. Everyday household behaviours depend on deeply held motivations, social norms and psychological predilections (Barr 2003; Head and Muir 2007b).

Likewise, we do not automatically assume that households should bear the burden of responsibility for transformation. In various chapters lively debate emerges over whether, for instance, manufacturers ought to bear the responsibility of disposal and recycling – from nappies (Chapter 1) to mattresses (Chapter 8) – or whether retailers ought to be more transparent with the energy and water use impacts of various goods on sale, such as televisions and computers (Chapter 13). Rather than looking for patterns of compliance with government sustainability objectives, we look instead at prosaic actions, and motivations that might be pro-environmental, or more-than-environmental.

Climate Talk – and Action?

A second way in which households are placed in climate change and sustainability debates is in relation to scientific knowledge and ideas – the manner in which information flow connects with or influences practices. Research is increasingly concluding that programmes informed by science encounter and generate scepticism and distrust as a matter of course – not as a result of poor communication or misunderstandings, but as a result of public assessments of institutions and clashes of divergent rationalities (Hinchliffe 1996). People may accept climate change science, but not act on it because climate change may be unthinkable within the confines of everyday life.

Relevant here are examples of science communication and behavioural change research. Scholars have sought to better understand individual response to governmental policy changes (e.g. Barr 2007), and what makes government-sponsored education campaigns effective or otherwise (Lorenzoni et al. 2007). Factors preventing or facilitating behaviour change are becoming clearer: for instance that fear fails to catalyse action (Hulme 2007), and that affective and emotional registers are as important as factual information in engaging with an issue as opaque as climate change (O'Neill and Hulme 2009). Emotions and the senses (sight, hearing, smell, touch and taste) also filter everyday practices such as eating (Chapter 2), wearing clothes (Chapter 3), toileting (Chapter 6), cleaning (Chapter 7) and driving (Chapter 10). Sustainability is embedded in daily bodily rhythms, senses and emotions irrespective of what courses of action might appear 'rational'. How 'households' are conceptualized in climate change adaptation talk, and whether this relates to the way people live and make homes, is a key tension running throughout the book.

Structures and Regulations

Third, broader economic and social trends provide extra complexity and contradiction. At the broadest level, only a small proportion of technological innovation is geared towards reducing energy or resource consumption, especially around industries such as electronics and information technology – innovation is driven instead by the profit motive and the saleability of new features and functions (Røpke 2012). This, in turn, is structurally linked to wealth – to those who can afford to purchase the newest, most advanced home products.

Notwithstanding the myriad subtleties discussed throughout this book, the strongest predictor of carbon footprint/greenhouse gas emissions is affluence, at both the macro and the household scale (ACF 2007). The best way to reduce your environmental impact is to be poor, as economic activity is strongly coupled to fossil fuel use. The rich pollute more through more and bigger houses; more food wastage; more consumption generally. Yet the rich and well-educated may be among the strongest advocates of 'green' practices: recycling, composting, buying organic food, taking reusable bags to the supermarket. They may also be leaders in buying still-expensive hybrid cars, solar electricity panels and green energy. Here lies a deeper dilemma: between 'green' as a means of distinction, to accumulate cultural capital (Askew and McGuirk 2004), and the pro-environmental motivations underpinning action. It is a dilemma borne out in various chapters of the book – from clothing (Chapter 3) to flying (Chapter 11) – where we report from our ethnographic and survey work on the contradictions of middle-class green consumerism.

Related to this intersection with broader economic processes is the volatile nature of the global economy – currently experiencing financial and sovereign debt crises in the United States and Europe. It is no coincidence that the only time in recent decades where carbon emissions actually fell globally was in 2009, when global economic growth collapsed in the wake of the financial crisis (Climate Commission 2012). At the centre of this were households, who, research has shown, responded to global economic crisis by reducing driving, becoming more frugal shoppers and spending more time at home as opposed to going out or on holidays (Leinwand et al. 2008). All of these translated into reduced emissions. Perversely, continued volatility may be the best thing for immediate climate change mitigation efforts.

The complication at the scale of households is, of course, that volatility manifests in widespread unemployment and threats to livelihoods. Even in the highest-carbon emitting countries of Europe and North America it cannot be assumed that households can respond to sustainability initiatives from a state of prosperity or stability (Leichenko et al. 2010). The household may be one scale where meaningful moves to reduce carbon emissions can be made, yet it is also at the scale of the household that economic risk is simultaneously, and most profoundly, experienced (Aalbers 2008). One outcome may be reduced household consumption as families tighten their belts. Another is that green consumerism, which grew during the boom times of the 2000s (Beard 2008), becomes more elitist. Nevertheless, as we shall see throughout the book, much of the work of sustainability is already being done by those with least capacity to pay – for example elderly people who exercise frugality and thrift as a matter of course, with memories of making do amidst world wars and rationing. Wealth is far from the only precursor to sustainable acts.

Beyond overall wealth as a key variable, other structural and material factors intersect with dilemmas of household sustainability. Both energy and CO_2 are strongly linked to type of dwelling, tenure, household composition and rural/urban location (Druckman and Jackson 2009). Although low-density urban form has worsened car dependency (Chapter 10), there are other benefits, as Chapter 16 shows in relation to gardens. Some argue that the most critical contributions to household carbon footprint stem from life-course transitions such as having a baby (Chapter 1) or getting divorced (Yu and Liu 2007, Murtaugh and Schlax 2009). But policy imperatives that might stem from some such studies – for example, having fewer children, or avoiding getting divorced – are likely to have missed the point. Can it be assumed that households will willingly change behaviours – and even family composition – in line with government objectives? This plays out in the book in various ways, in relation to decisions to have children (Chapter 1), to live in extended family households rather than traditional nuclear structures (Chapter 13), and whether to downsize or consume less when retiring from the workforce (Chapter 18).

Regulatory frameworks also vary greatly – as do relative contributions of nations to global warming. Only 37 countries had targets under the Kyoto Protocol (although 90 per cent made 'pledges') (Climate Commission 2012), and the per capita contributions of, for instance, Australians (27.3 tonnes of carbon dioxide equivalent (CO_2-e) per person, per annum) and Americans (23.4 CO_2-e) are substantially higher than most other OECD nations (Japan 10.5 CO_2-e; European Union 10.3 CO_2-e) (Climate Commission 2012). Comparing household sustainability dilemmas across such countries is not straightforward. Carbon pricing has the potential to even out comparison, but it too is being introduced in fits and starts internationally – in some countries at the city (Japan) or state (United States) scales, in other jurisdictions as multilateral initiatives (e.g. the European Union) (Climate Commission 2012). Potential improvements in emissions may have less to do with household practices than with transitions in regulatory regimes, in supply chains or in energy generation methods nationally: when, for instance, national power companies switch fuel sources (from coal to gas) (Druckman and Jackson 2009). Another is the structure of the food supply chain – where for example much food wastage occurs because of the nature of the legal agreements between farmers and supermarkets (UN FAO 2011). Some things are out of the immediate hands of households.

Nevertheless energy labelling schemes can assist households in making purchasing decisions. The most commonly labelled appliances are refrigerators, freezers and air conditioners – though in some countries labels are also available on rice cookers, boilers, lighting products and washing machines (Boström and Klintman 2011). There are wide variations nationally (Harrington and Damnics 2004). Not all labelling schemes help, either. Best-before dates on food products are a rational response to public health concerns over freshness, and have helped reduce the risk of gastrointestinal illness, but nevertheless generate huge amounts of waste, encouraging those who can afford to waste food to throw away what might be edible items (Chapter 12).

THINGS

> The home place is full of ordinary objects. We know them through use; we do not attend to them as we do to works of art. They are almost a part of ourselves, too close to be seen. (Tuan 1977, 144)

However solid the physical dwelling, the home is in one sense nothing more than a membrane through which energy and stuff flows (Biehler and Simon 2011), in what Fine and Leopold (1993) called 'systems of provision'. For commodities to come to be as they are requires routes of manufacture, distribution and consumption, architectures, manufacturing processes, infrastructures, environmental transformations and cultural meanings. Such practices and processes are not neutral; rather they generate ethical complexities that require interpretation within normative environmental and social justice parameters (Castree 2004). Throughout chapters in the book we make attempts to 'follow the things' (Cook 2006) on their journeys from raw materials to production, distribution and home consumption.

What transpires is that those journeys are often complex and opaque, making it very difficult to ascertain the interrelationships between products consumed and downstream environmental impacts (Lenzen et al. 2012). In just one study – of the lifecycle assessment of disposable nappies (Chapter 1) – researchers had to take into account land area required to grow trees to pulp into cellulose (the absorbent part of nappies), which in turn required consideration of varying forestry practices, which then led to consideration of run-off impacts, which in turn impacted stream flows, which can be detrimental to freshwater ecosystems (O'Brien et al. 2009). Put another way, the nappy on a child's bottom is linked causally – if in complex ways – to deteriorating fish spawning ecosystems in a stream near a pine forest somewhere unknown. It is also likely linked to soil pollution in a landfill site on the edge of a municipality somewhere else, where, it is estimated, the nappy will take 200 to 500 years to break down.

Some systems of provision are very fixed, and some are fluid. Where they are fixed, any changes that a household makes may be limited unless connected to larger scale infrastructural and technological change (Lawrence and McManus 2008). Where they are fluid, households may be able to contest wider patterns of consumerism through bargaining networks and informal sharing with friends, relatives and neighbours. Here details of the materiality of the things – and the social uses to which they are put, such as sharing (Belk 2010) – matter enormously.

The material spaces of the household, and the things in them, can be structured by human activity, but also have agency in their own right. Home designs with poor passive solar design ratchet up energy use in heating and cooling (Chapter 5), as do appliances with stand-by functions, and washing machines that default to warm water washes (Chapter 7). Tablets, smartphones and Wi-Fi technology combine to make 'always-on' the norm for home internet connectivity, using energy 24-hours daily (Chapter 13). Refrigerators stretch the period in which we can safely eat food (and leftovers), but also encourage us to buy (and waste) more food (Chapter 12). Several of the things in the house cooperate with our wishes and go largely unnoticed – mattresses (Chapter 8), fridges (Chapter 12), hot water heaters (Chapter 15) – until they degrade, break down, or decide to become entirely uncooperative.

At various points throughout the book we pursue such things about the home, taking seriously the 'vitality of matter and the lively powers of material formations' (Bennett 2010, vii). This is not meant to transfer responsibility for acting from humans onto the 'things' we consume (or their design – as if mere improvements in technology were enough), but rather to take seriously 'the capacity of things – edibles, commodities, storms, metals – not only to impede or block the will and designs of humans but also to act as quasi agents or forces with trajectories, propensities, or tendencies of their own' (Bennett 2010, viii). Gay Hawkins argues this case powerfully in relation to household waste. Any cultural understanding of waste needs to go beyond the fantasy of human control over non-human others:

> As one of our most everyday habits, disposal depends on a particular kind of blindness that helps us *not* see, *not* acknowledge the things we want to be free of. To throw things away is to subordinate objects to human action, it is to construct a world in which we think we have dominion. This doesn't just deny the persistent force of objects as material presence, it also denies the ways in which we stay enmeshed with rubbishy things whether we like it or not. (Hawkins 2006, 80)

Seemingly prosaic acts – starting a worm farm or compost bin, recycling paper and other materials for children's school art projects – interrupt flows of material stuff in and out of the household, forging a 'different means of engagement with waste' (Hawkins 2006, 128; see also Hetherington 2004; Moore 2012). They also reconfigure relationships with worms, microbes and children. How we engineer and design goods for their reuse as well as intended purpose is critical – as is a stronger sense of stewardship in relation to the things in our lives (Lane and Watson 2012). In this vein, Gregson et al. (2007, 683) tracked practices of storing and reusing, arguing that

> whilst people certainly did get rid of consumer objects via the waste stream, they also went to considerable lengths to pass things on, hand them around, and sell them, and – just as often – quietly forgot about them, letting them linger around in backstage areas such as garages, lofts, sheds, and cellars, as well as in cupboards and drawers.

Such 'stockpiling', we shall see, features in a number of the chapters in this book, in relation to clothes (Chapter 3), furniture (Chapter 8), electrical equipment, books and CDs/DVDs (Chapter 13) and phones (Chapter 14).

We ask how easy or difficult it might be to reconfigure connections between people, things and material spaces – and what are the sustainability implications of such reconfigurations. How is sharing of appliances assisted or hindered by house design? We shall see how families turn garages into bulk storage zones; how bedrooms become spaces of private recluse; how open plan designs – problematic for their energy-sucking qualities when heated or cooled – also enable communal meals; how lower floors are turned into granny flats – each in turn contributing to altered per capita energy and water use. But are there limits? How easily are televisions and cars shared? How such things play out is in large part a function of cultural proclivities towards communal family life, comfort, cleanliness, privacy and independence, but also a function of underlying infrastructure provision (roads, recycling schemes, telecommunications infrastructure). In all this, cars, televisions, wires, bins, home layouts and kitchens are animate, lively things.

EVERYDAY PRACTICE

Most incentive and education programmes pay little attention to the ways household energy, water and other resource consumption practices are part of the rituals, rhythms, emotions, habits and routines of everyday life (Shove 2003; Gregson et al. 2007). Cultural norms shape household consumption in complex and uneven ways (Lorenzoni et al. 2007). Sustainability campaigns normally fail to appeal to or appreciate the emotional meanings attached to material possessions (Allon and Sofoulis 2006; Hobson 2008) or home spaces (Blunt and Dowling 2006) – or because of 'counterintentional' habitual behaviour (Maréchal 2010). People fly around the world for business and holidays, and to visit loved ones, even when they know about the intense environmental impacts of aeroplane travel (Chapter 11). So too austerity, hoarding, sharing and charity donations – all cultural practices with implications for reducing consumption – require analysis in specific social settings to ascertain motivations and meanings.

Increasingly sophisticated waste management systems now enrol households in recycling, composting, sorting and disposing, but what constitutes 'waste' in households is far from straightforward, and official schemes only capture a proportion of the potentially recyclable waste stream. The corporatization and technological sophistication of household waste schemes contrasts with everyday practices of disposal, reuse and recycling – many of which are highly informal and operate outside official schemes (Lane et al. 2009). An older generation who grew up with the frugality of a depression and wars might save and reuse glass jars, tin cans and bits of string, items that younger family members rid, as 'waste', or send off as recyclable things through official schemes, as a matter of course (Chapter 19).

Inside the house we encounter norms of cleanliness, for both human bodies and their clothes, which embed increasing levels of water consumption in the bathroom and laundry (Shove 2003; Troy et al. 2005; Allon and Sofoulis 2006). Is showering a means to hygiene (Chapter 7) or a 'leisure activity' (Chapter 15)? How often should clothes be washed (Chapter 3)? Teenagers may have four changes of clothing and more than one shower a day, for example as they exercise, attend university, go to their part-time job, and then go out at night (Sofoulis 2005). The particular dirt of each context has to be removed by washing from both bodies and clothes. Chapter 7 explores such issues and the measures people take to avoid being sweaty or dirty, while Chapter 6 explores the confronting extension of this: how we relate to our own (and others') faeces, through toileting practices and norms. There are clear implications in these examples for water and energy consumption. Other forms of household 'waste' are less likely to be experienced in terms of human disgust, but are stockpiled, stored or displayed in wardrobes, attics and basements: 'the detritus of urban life ... doesn't destabilize the self. It just hangs around largely ignored' (Hawkins 2006, 3). Collecting nevertheless opens up awareness for sharing and an afterlife for things.

Throughout the book we identify frictions and points of traction to help think about how different elements of governance, materiality and practice interact in the context of the household. Technology, cultural meaning and social practice converge in a variety of ways that are both resistant and amenable to transformation. We also draw on Shove's (2003) metaphor of the ratchet to discuss the role of tools and technologies in the making and remaking of everyday household practices. She illustrates how changing social norms, say in terms of cleanliness and washing clothes, may counteract efficiency improvements within systems of provision. The recent permeation of information and communication technologies (ICT) into households is a different sort of example: a 'new round of household electrification, comparable to earlier rounds that also led to higher electricity consumption' (Røpke et al. 2010, 1764). The uptake of ICT and Wi-Fi ratchets up emissions. Nevertheless changing use of computers and televisions opens up spaces of traction, for instance through reduced material production of CDs, DVDs and books (Chapter 13). In many ways friction and traction are two sides of the same coin, but we use them here to trace less and more sustainable pathways respectively. So friction may involve pathways of resistance to more sustainable outcomes, or contradictory practices that entrench problems. Traction can result from the de-routinizing of previous practices – when, for instance, families move houses and old habits are – temporarily at least – interrupted (Maréchal 2010). The term traction also helps identify useful points of intervention: policies, key players, levers, intermediaries or translators, both human and not.

MEASUREMENTS AND METHODS

Before we begin our journey through the home, some words are needed on measurements and methods used in research on household sustainability (and in our own survey and ethnographic projects). At various points we draw on published, peer-reviewed analysis of the environmental impacts of forms of production, and activities such as flying, driving and dying. We also draw heavily on our own suite of quantitative survey and qualitative ethnographic research with households.

Methods of assessing environmental impacts vary – and indeed the devil in the detail for household sustainability is magnified by variations in the methods of accounting used. Most commonly we draw on previous studies of accounting energy, water and carbon emissions, and on lifecycle assessment (a 'cradle-to-grave' study of a product's environmental impacts, from raw material extraction through manufacture, distribution, home consumption, use, repair and disposal or recycling). To a lesser extent we also cite results from ecological footprint analysis (a measure, *global hectares*, or gha, in total land area, of the total demand on the earth's ecosystem generated in order to support a lifestyle).

Each method has its weaknesses, hampering interpretation of sustainability dilemmas. Lifecycle assessment, for instance, is based on calculations of averages, often within bounds of a spectrum of conservative to extreme estimates. Lifecycle assessment provides some means of comparison between products, but struggles to capture wild differences in the *manner* in which people use a product. This becomes relevant, for instance, in the chapter on screens (Chapter 13), where the debate about digital versus hard copy consumption of reading material, music, films and television shows depends a great deal on exactly how many books, CDs or DVDs are being replaced by an eReader or laptop. Intensive use of a high-carbon embedded product by one person might cast an entirely different light on that product's average lifecycle impact.

Where we refer to statistics on carbon emissions, we tend to use the carbon-equivalence figure (expressed as tonnes of CO_2-e). Visualizing tonnes of CO_2-e is not at all easy: unlike solid materials, carbon emissions are difficult to gauge in quantity. For what it is worth, Wright et al. (2009) provide a useful set of comparisons that are more easily imagined: one standard plastic garbage bag, if filled with carbon dioxide, would contain about 100 grams of it. Or put differently, one tonne of CO_2-e can be imagined as 10 000 garbage bags filled with carbon dioxide.

There are difficulties, too, in tracking carbon emissions accurately across a population, let alone for individual products or actions. CO_2 embedded in consumption, for instance, is calculated using methods of environmental input-output modelling that are data and resource intensive – complicated enough even if the required data is available within one country; wildly complex if the product in question is made elsewhere, or made using components from several countries (Druckman and Jackson 2009, 2067). The result is that some things can be more reliably assessed for their carbon emissions than others: CO_2 emissions resulting from direct household fuel use, flying or driving is relatively straightforward, whereas estimating the embedded emissions in a television or a computer is a practical nightmare.

Care must also be taken in interpreting singular calculations of the lifecycle impacts or carbon emissions embedded in individual products, as if their 'report card' rested solely on their separate existence. As discussed above, a relational view of household sustainability dilemmas seeks to interpret humans and various non-human things in networks of distributive agency within and beyond the family home. What this means is that all products and activities assessed need to be held in tension with the calculations of all others within a household, and contextualized within the practices and dynamics of household life. Thus it is possible that changed consumption practices eliminating carbon emissions in one area of household life may be replaced with others that ramp up emissions. Saved money from one thing is transferred to another, with negligible overall carbon emissions savings. Druckman and Jackson (2009, 2068) illustrate with the example that 'when respondents in a survey were asked how they would spend any savings accrued from lower energy bills, the most common single answer was "an overseas holiday involving air travel"'. Such rebound or backfire effects are highly unpredictable and context-dependent.

Supplementing our use of published statistics and research findings is our own data, generated via successive large research projects in our research centre. First was a large household survey conducted in Australia in July 2009 as part of a project drawing on mixed methods to explore climate change in everyday life. A survey entitled 'Tough Times? Green Times? A survey of the issues important to households' was posted to a random sample of 11 555 households, inviting an adult familiar with the running of the household to participate. The survey comprised both open and closed questions, and addressed socio-economic characteristics, household consumption practices, everyday objects, activities and practices, and judgements of climate change.

The sample size reflected the need to generate approximately 200 usable survey returns from each income quintile range in the total population, to permit statistical testing, based on an expected overall return rate of 10 per cent. The actual return rate was 12.67 per cent (n: 1 465). Respondents were statistically representative of the total population sampled; and returns were therefore not weighted. Quantitative analysis was conducted using Statistical Software Package for Social Sciences (SPSS) and included a variety of statistical tests and computations (see Waitt et al. 2012b for further detail). When we refer to 'our survey results' throughout the book, it is from this large quantitative activity that statistics are drawn.

Throughout the book we also draw on qualitative ethnographic work conducted over a decade on a series of interrelated projects. These projects revealed the nuances and contours of households and the dilemmas of sustainability in everyday life. One major project was an in-depth longitudinal ethnography of households in Australia. Interviews and observations were conducted in 16 households every three to four months between January 2010 and November 2011. Discussions covered a range of everyday practices such as shopping and laundry. Households volunteered to participate in the longitudinal research following the above postal survey on household sustainability. To an extent, the recruitment process meant that participating households were likely to be engaged with environmental issues and to be participating in recycling, water conservation and other mainstream environmental practices. All were concerned with, and practised, some degree of resource stewardship. All were concerned with environmental degradation, and some were engaged in local sustainability initiatives. Our household participants represent a range of occupations, incomes, educational attainments, sexual and ethnic identifications and household compositions.

Other samples were derived as complements to this main ethnography, on which we draw here for more specialist purposes. One was extended family households (interviews conducted with 17 participants from ten households) living in owner-occupied, detached dwellings in suburban settings. This sample provided insights on the unique challenges faced by households of larger average size, combining family units, as well as intergenerational relationships (see Klocker et al. 2012 for further detail and explanation). Another was a sample of six environmentally committed households participating in, and recruited through, Sustainable Illawarra's Super Challenge Program. These households were the focus of a specific sub-project on gender and time, and who does the work of sustainability in the household. A mixed method approach included a combination of home tours, diaries, photography, video camera logs, time charts and in-depth interviews (see Organo et al. 2012 for further detail). At various points throughout the book we also draw on more specialist and/or allied ethnographic projects that we or our postgraduate students have pursued: on solar hot water (in collaboration with CSIRO's Peter Osman), water tanks (Moy 2012), backyards (Head and Muir, 2007a), clothes (Gibson and Stanes 2010), cars (Waitt and Harada 2012), tomatoes (Roggeveen, 2012), kangaroo meat (Appleby 2010) and wheat (Head et al. 2012). It is to such seemingly ordinary, everyday things in the household that we now turn.

1. Having a baby

Are you planning on having a baby? As any parent will truly know, it is one of life's most significant decisions. Having children also brings with it confronting sustainability dilemmas.

At the outset, having a baby is directly linked to overall human population size, inviting consideration of the total consequences of an expanded population for the environment. When families decide to have children they contribute to birth rates, and thus call on the earth to support another human life from its stock of physical resources. This is not so much an issue if death rates are also high – as was the case for humans up until the Industrial Revolution. But since the advent of widespread health care, urban sanitation and improved practices in midwifery and disease prevention, humans have found ways to die less frequently, to stay alive longer. As a consequence, overall human population growth has soared, in turn amplifying demands for resources. As Hall et al. (1994, 506) put it, 'the ultimate environmental impact occurs with the birth of each new human being, for a whole suite of production and consumption activities commence with that birth'.

What exactly is the overall environmental magnitude of the decision to have a baby?

Hall et al. (1994, 506) developed an environmental impact statement 'for the birth of one baby in the 1990s in the United States', accounting for a hundred environmental impacts that one person born at the time of the study would cause over an expected lifetime (grouped under the headings 'waste generation', 'mineral consumption', 'energy consumption', 'food consumption', 'ecosystem alteration'). The results are stark: an American born at the time of their study

> would produce in a lifetime about one million kilograms of atmospheric wastes, ten million kilograms of liquid wastes, and one million kilograms of solid wastes... will consume 700 thousand kilograms of minerals, and 24 billion BTU's of energy, which is equivalent to 4 thousand barrels of oil. In a lifetime, an average American will eat 25 thousand kilograms of major plant foods and 28 thousand kilograms of animal products, provided in part by slaughtering two thousand animals. (Hall et al. 1994, 509–510)

Murtaugh and Schlax (2009, 14) updated calculation of the carbon legacies of reproduction; they made the point that 'a person is responsible for the carbon emissions of his [sic] descendants', across several generations – and not just for their immediate children. The result, they estimate, is that each new child adds a cumulative 5.7 times the lifetime emissions of their mother.

Notwithstanding the power of such statistics to force contemplation of our total environmental impact, there are two problems with this kind of macro-scale thinking. The first is the recommendation in both studies that the best thing households can do to reduce their environmental burden is to refrain from having children. Hall et al. (1994, 523) argue that 'the most effective way an individual can protect the global environment, and hence protect the well-being of all living people, is to abstain from creating another human'. But can we reasonably expect that households forgo having children on purely environmental grounds? That seems an unlikely choice to be made voluntarily. Certainly, factors such as cost, time and social norms govern family size. As Hall et al. (1994) themselves acknowledge, families have children for all manner of reasons: to support parents when old, for company, for emotional bonding, to contribute altruistically to the societal reproduction of future generations.

The second problem is that overall population growth is compounded by the per capita consumption of resources by humans. What matters is not so much that you exist, but how much of the earth's resources are required to support your lifestyle (and those of your children). While the overall human population has escalated since the Industrial Revolution, so too has per capita consumption. Arguably the drivers of global population growth have been addressed through improved global health and family planning. Consumption, meanwhile, spirals unabated. What people *do* once they have babies is critical.

MATERIAL PRACTICES OF PARENTING

Having a baby can be a shock to the system. It was no surprise, then, that in our ethnographic research those families who recently had children cited that event as a significant factor in changing practices. There were some predictable and straightforward increases in resource consumption. In our survey, of those who recently had a baby, 87 per cent had increased energy consumption as a result, and 91 per cent had increased meat consumption. Not surprisingly, some 95 per cent also recorded a decrease in social outings. Influences on transport use were less clear cut: of those whose transport use had changed after having a baby, 58 per cent had increased transport use, but 42 per cent had decreased transport use. As Justin Spinney (2012, 1) has argued, 'parenting imposes a new suite of materialities, affective capacities, time constraints and practices, which in turn inform daily travel choices'.

Beyond the sheer presence of another human being are further baby-related practices – and sustainability dilemmas. The typical shopping list for affluent new parents includes stocks of nappies (reusable cloth or disposable), a car safety seat, a pram, a cot (with linen), a baby change table, sterilizing equipment, bottles and plastic crockery, a high chair, a rocker, toys, clothes, wraps, and blankets. With this surge of consumption in mind, a number of environmental non-government organizations and government programmes have sought to promote more sustainable options for families with newborns.

Breastfeeding is, for example, 'a no brainer', according to Treehugger.com. Beyond clinically proven health benefits and the importance for mother–child bonding, breastfeeding minimizes environmental impacts (other than food production needed to feed the mother), doing away with the need for plastic bottles, teats, sterilizing equipment, and the water and energy used to clean them. Nevertheless, as Treehugger.com wryly point out, 'in our commerce-driven society there are products for everything, and breastfeeding is no exception'. There are 'greener' breast pads (reusable organic cotton or wool felt pads rather than disposable) and organic nipple creams. Fair-trade organic infant formula is available, though not universally (and it is expensive).

For every other aspect of parenting newborn babies there are similar prescriptions: for solid foods choose organic and avoid buying food in jars that maximize glass and container production; make your own baby meals and freeze large batches; search for organic baby clothing; consider plain olive oil instead of petrochemical moisturizers; use more gentle cleaning agents around the house rather than high chemical-input disinfectants. Having children has, in short, become another means to practise green consumer-citizenship.

Further complexities stem from the systems of provision required to produce and distribute each of these baby-related products, such as the political economy of food production (Chapter 2). The market for organic baby foods has grown considerably in the past decade but not only due to environmental concerns: parents want to feed their children the 'purest' food without risk of pesticide and fertilizer residues (Hughner et al. 2007). This contrasts with the imperative for parents to cook their own food, reducing packaging waste, but increasing the time burden (usually on women) compared with ready-to-eat foods.

Another vivid example is that of baby clothing. The market for organic hemp, cotton, bamboo and wool fabrics has expanded most quickly for children's clothing – again linked to parental concern for welfare of their children (as well as for the environment). But as discussed in Chapter 3, the sustainability ethics of the 'eco-fashion' movement are far from straightforward. Such fabrics also require resource and energy inputs. Eco-branded clothes are likely to be purchased by wealthier families and they are a means to accumulate cultural capital, suggesting the motivation to 'feel good' about clothing consumption is as important as tangible reductions in environmental impact.

An alternative is sharing and handing-down used clothing (Clarke 2000). It is common practice nowadays for boxes of clothes, blankets and toys to 'do the rounds' through networks of family, mothers' groups and friends (Waight 2011). The mobility of such things extends commodity use value and limits resource use, but also marks the end of a particular phase of parenting, and 'makes new mothers (and generations) through the passage of baby-related and children's things' (Gregson 2007, 93). Nevertheless, there are limits and contradictions: advice to reuse cots, furniture and prams rubs against norms of 'newness' and conflicting rationales, and well-intentioned people are warded off reusing cots and car seats because of safety concerns or the risk of Sudden Infant Death Syndrome (reported to be higher in reused mattresses/cots) (see Chapter 8).

THE BIG NAPPY DEBATE

Nowhere are the sustainability dilemmas of new parenting as problematic as with nappies. As the *Sydney Morning Herald* (*SMH* 2007) reported, 'Nothing quite polarizes sentiment or induces more outrage among the environmentally-conscious than the question of whether you use (or used) cloth nappies or chose disposables'.

Figures vary between countries, but overall, the numbers are arresting: before toilet training (usually between 2 and 3 years) an average baby uses 6000 nappies, produces 91 kg of faeces and 458 kg of urine. Forty-nine million disposable nappies are used per day in the United States; 2.2 million in Australia, 6.7 million in Japan, and 9 million in the UK (Treehugger.com). In Australia and the UK 95 per cent of babies wear disposable nappies (Aumônier et al. 2008; O'Brien et al. 2009). Australians discard nearly one billion used disposable nappies annually (CSIRO, quoted in SMH 2007), Americans discard over 10 billion. A UK Environmental Agency report found that the nappies required for one child until toilet trained (averaging at two and a half) are 'roughly comparable with driving a car between 1300 and 2200 miles' (quoted in SMH 2007). Non-biodegradable disposable nappies take an estimated 200–500 years to break down, and are according to some reports the single biggest category of item dumped in landfill.

On such figures alone it would seem straightforward to suggest that reusable cloth nappies present the more sustainable, if more time-consuming, alternative.

Coming to reliable conclusions is, however, remarkably difficult. In part this is because published lifecycle assessments of different types of nappies are frequently funded by companies and industry associations with vested interests – disposable nappy manufacturers such as Procter and Gamble, or industry associations lobbying on behalf of commercial laundering nappy services.

An early and much-cited Canadian study, financed by Procter and Gamble, modelled consumption of energy, raw materials and water, atmospheric emissions and wastes, and found that cloth nappies consumed more water and produced more waterborne wastes, whereas disposable nappies consumed more raw materials and produced more solid wastes (Vizcarra et al. 1994). Cloth nappies used 15 per cent more energy per baby per week, and double the water; those home laundering cloth nappies used half as much energy again, and 70 per cent more water than those using commercial laundering services. All up, the laundering of cloth nappies accounted for between 81 and 93 per cent of their energy consumption. On the other hand, disposables used seven times the raw materials, while for cloth nappies around half the raw materials were used in the manufacture of detergent and bleach needed to clean them. Atmospheric emissions from both were about equal – though for cloth nappies up to 83 per cent of emissions resulted from laundering (the most highly variable element depending on home versus commercial laundering, wash temperature and dryer versus line-dried). In the researchers' words, 'neither diaper system can be considered absolutely superior environmentally' (Vizcarra et al. 1994, 1707).

Assumptions made prior to modelling can, however, make an enormous difference to observed outcomes. Are used cloth nappies recycled as rags or disposed to landfill? Are cloth nappies home laundered or are commercial laundering services used? At which water temperature are they washed? Are they pre-soaked or 'dry pailed' (stored in a bucket for later washing)? Results can vary enormously depending on whether front-loading or top-loading washing machines are used. A 2009 Australian study concluded that 'reusable nappies washed in a top-loading washing machine were found to use more water than disposable nappies or reusable nappies washed in a more efficient, front-loading machine, regardless of the number or mass of the nappies used' (O'Brien et al. 2009, 6). The use of electric dryers as opposed to line-drying also dramatically skews results. In Canada, 68 per cent used electric dryers and only 32 per cent used line-drying (Vizcarra et al. 1994). The most reliable UK study assumed that 25 per cent of nappies were tumble-dried (Aumônier et al. 2008) whereas in Australia a similar study assumed that all cloth nappies were line dried (not an unreasonable assumption given that country's drier, warmer climate). Most industry-funded studies do not take into account runoff issues associated with the production of raw materials needed to make nappies. When O'Brien et al. (2009, 8) did, they found that 'water resource depletion associated with the softwood production for disposable nappies was higher than for the cotton growing stage of home-washed nappies'. Such differences can make a significant impact on lifecycle assessment – such that one or another type of nappy can appear superior over the other in terms of total environmental footprint.

The comparative magnitude of different sorts of impacts is opaque. Weisbrod and Van Hoof (2012) argued that sourcing and production of disposable nappy materials accounted for 84 per cent of all non-renewable energy uses and 64 per cent of global warming potential – while disposal was a small contributor (1–12 per cent) of overall environmental impacts. Yet this invites downplaying waste disposal predicaments, which are still significant, and contradicts assessments elsewhere that 'the problem with disposables is the amount of landfill they produce... and the fact that toxic contents are not treated via the sewerage system' (SMH 2007). O'Brien et al. (2009) found that commercially washed nappies saved much more water than home-washed reusable nappies. On the other hand, commercially washed reusable nappies had a shorter lifespan, and thus water use required to support necessary cotton production was increased. No wonder people are confused.

The most significant impacts associated with cloth nappies are less visible – bound up in post-purchase laundering rather than their physical manufacture – but at least within the control of consumers. Disposable nappies on the other hand enrol complex pre-purchase commodity chains in production, over which the consumer has little influence. Taking into account the whole lifetime and journey of cloth versus disposable nappies involves comparing two products that are perhaps not, ontologically, the same kind of thing.

Adding to the complexity – and adding fuel to the sometimes heated nappy debate – is that constant improvements and changes are made to nappy types and 'systems'. Better-fitting, shaped 'modern cloth nappies' are now more common; they require less frequent changing and can be used again for second children. Washing machines and dryers are also more efficient. Manufacture of disposable nappies has become less energy and resource intensive. Biodegradable disposable nappies have become available, but are higher-priced compared with other disposables, and only work if the used nappies are exposed to oxygen and composting bacteria (as opposed to wrapped in plastic bags and disposed through household rubbish collection destined for landfill; see also Chapter 9). They are also are harder to find.

Whatever the actual merits of the type of nappies used, there is no doubt that it has become an emotive issue. Nappy use elicits heated debate arguably more than any other dilemma in this book. Comments made online in response to a 2007 *Sydney Morning Herald* article on the sustainability issues surrounding nappies show how deeply shame, guilt, anger and pride infuse the debate.

One refrain is to direct shame towards those making and using disposables, from the standpoint of (perceived) superior environmental citizenship:

It's a crime to pollute generally and yet the manufacturers of disposable nappies are allowed to get away with pollution on a massive scale with their non-biodegradable environmentally destructive products. Where is the sense in using a product on your baby which is going to directly impact on the quality of life they will have in a polluted toxic world? Shame on parents who choose the non-environmentally friendly option and shame on the retailers in Australia who stock those brands over the biodegradable ones. (*SMH*, Nappy Head, 17 April 2007)

Another variant fuses concerns about cleanliness, and emotions of disgust towards human waste (see also Chapters 6 and 7):

I am vehemently opposed to disposable diapers. Those who use them seem to think it is acceptable to leave them anywhere they like. They contain sewage which should be disposed of via the sewerage system. Disposables cannot be disposed of in this way. Cloth nappies should be emptied into the toilet and soaked in disinfectant and later washed hygienicly [sic] using an appropriate detergent. (*SMH*, Claire, 17 April 2007)

Using cloth nappies can be a point of difference and pride, a means to feel good, and also a trigger for further vernacular environmental adaptations:

I have a 14 month baby and I use cloth nappies. For me it is an environmental reason and also a cost factor... In Australia I dry my cloth outside in the sun. Not an electric dryer. I put the water out on the garden... I don't add to landfill and I wash number 2s down the sewerage system. My babies poo is not put into a plastic bag and then into a bin. I use a cotton wash cloth to clean her up. My garbage is down to 1 medium size bag a week... The amount of time that I use to clean nappies is not long at all but I feel great about myself. When my baby grows up and asks me what did I use cloth nappies or disposables, I will tell her proudly I used cloth. (*SMH*, Sarah, 17 April 2007)

Others are quick to defend disposables:

As a parent, disposables are one of the greatest inventions of the 20th century. There are so many other ways to reduce environmental impact – why put yourself through the pain of the cloth nappy routine when, as has been shown, the environmental impact is almost identical. (*SMH*, SP, 17 April 2007)

Some disposable nappy users measured environmental costs against broader constraints: 'I turn off my lights... I separate my rubbish, quite meticulously, however I used disposable nappies and I refused to feel like a criminal for it!' (*SMH*, Kathy, 17 April 2007). In some cases, balance was sought between comfort for the baby and environmental gains achieved elsewhere in the house:

For us, at the end of the day, our baby was more comfortable in disposables. A baby wearing cloth nappies may as well just be peeing in their clothes... So given the fact that the latest theory is that the environmental impact of either type of nappy is roughly the same, I'll choose to make my child comfortable, thanks very much. I'll choose the green option when it comes to other aspects of my household such as economical vehicles, light bulbs, water usage, etc. (*SMH*, Tim, 18 April 2007)

Some women felt that pro-environmental arguments raised by cloth nappy users unfairly placed extra workload onto women:

I would like to see the cloth advocates, with two parents working outside the home and more than one child, manage to keep up with the washing. The cloth vs disposable debate colludes in the continued oppression of women by implying that these choices are part of some 'female problem', rather than one of the choices that a family makes in managing its environmental footprint. Why are nappy users considered the font of all evil by people who reserve the right to drive whatever car they choose, whenever they like? (*SMH*, Old Bag, 17 April 2007)

In the mid to late 2000s in Australia an extra layer of complexity was added due to the prolonged drought. Drought alerted Australians to the prospects of climate change; metropolitan authorities imposed tough water restrictions and households looked to find new ways to save water (Chapter 4). Use of reusable cloth nappies now appeared villainous given the magnitude of water resource problems; and some felt guilty using reusable cloth nappies in a time of extreme water shortage. With such geographic and temporal variation, with scientific assessment of the full lifecycle and emissions footprint of nappy types inconclusive, and with vehement moral positioning among advocates of differing nappy types, a sense of stalemate pervades.

WHAT TO DO?

Are there other alternatives? One is to make your own reusable nappies based on sewing patterns available online. Elsewhere schemes have trialled methods to better manage disposable nappies: a nappy biowaste household collection and biodigesting project in Holland, and a disposable nappy recycling facility in the UK ('reclaimed materials will then be used to create a range of products including park benches, garden furniture, decking, bollards, railway sleepers, fencing, roof tiles and cardboard tubing' (Knowaste 2012)).

A more radical option is to attempt to reduce nappy use altogether, hastening toilet training readiness. The argument here is that Western norms of toileting and cleanliness (see also Chapter 6) and the advent of disposable nappies have delayed toilet training from 18 months to 2–3 years (Brazelton et al. 1999). Instead, an evolutionary and anthropological view of toilet training is being promoted, with inspiration taken from an influential 1977 paper by

Devries and Devries, in which Western norms are compared against those of the Digo of East Africa. Infant toileting practices are viewed as adaptive to survival and cultural values. In non-Western cultures 'infants can learn soon after birth and begin motor and toilet training in the first weeks of life. With a nurturant conditioning approach, night and day dryness is accomplished by 5 or 6 months' (Devries and Devries 1977, 170). Parenting experts, bloggers and paediatricians are now debating how to replace nappy use altogether, learning from the half of the world's population unable to afford either disposables or high energy and water use cleaning of cloth nappies (Dombroski 2012).

Nevertheless, within the paediatric literature and beyond, such techniques have been criticized for their risks of being overly coercive, and therefore psychologically damaging. Critics also list the risk of making parents who fail at it feel guilty (Brazelton et al. 1999), as well as being unrealistic for urban families – and women in particular who are increasingly returning to the workforce earlier, whether through choice or necessity. Such movements do, however, reverse the assumed rich-to-poor world flow of knowledge and expertise over hygiene and parenting, challenging the conflation of poverty with backwardness (Dombroski 2012).

Finally, as if all this complexity was not enough to deflate well-meaning parents, there is a wider debate about the effects of consumer culture on children, their exposure to voluminous television advertising encouraging wasteful consumption, and the movement towards simplicity parenting (which involves paring back children's 'stuff' in the home, including toys). As Pugh (2009, ix) put it, 'we are living through a spending boom that is unprecedented, and which is exacting a great price. Childhoods have become ever more commercialized, with hundreds of billions of dollars annually being spent on or by children in the United States alone'. Yet as Pugh (2009, x) goes on to argue, 'if consumer culture is the "enemy" of good parenting, why do so many parents invite the enemy into their homes?' The answers are, again, complex. Gregson (2007, 70) found that parents meaning well by their children 'invest heavily in the new' rather than give children used toys or hand-me-down clothes as gifts. In Pugh's (2009, x) words 'families [have] to invent their own versions of Christmas, Halloween, the Tooth Fairy, allowances, birthdays – each time adopting a particular stance toward the consumer culture that [is] banging on the door, peering in the windows, and sometimes climbing down the chimney to get in' (see also Chapter 17). Other social norms, such as those surrounding sleeping, also trigger replication and accumulation of 'things', with children accorded separate, individual bedrooms (and frequently duplicated in the case of divorced parents). Such are the travails of the modern family, making everyday parenting decisions in a moral landscape where commercial, environmental and societal norms collide.

2. Spaghetti bolognese

There is no more important household task than getting food on the table. Shopping for and preparing meals is a complex and relentless task in which decisions about cost, convenience, taste and health are juggled. Like other aspects of domestic labour (see Chapters 7, 12), it also remains a heavily gendered task. More of the financial provisioning is undertaken by men, with women doing more of the work of translating money into the family meal. When both do so under increasing time pressures, with or without the presence of young, hungry children, it is not surprising that the family meal table becomes a flashpoint. Yet the family meal retains iconic status as a symbol of sharing and communication around a common hearth.

What about sustainability concerns? The complex suite of materials and meanings around food create both potential for a sustainability focus, and competition with other household demands. Food is, according to one study, responsible for 29 per cent of the greenhouse gas emissions of a typical Australian household (Wright et al. 2009, 4). The dilemmas of sustainability around food are critical. Comparable national figures in Europe vary between 15 per cent and 28 per cent (Garnett 2011). Yet research suggests that food and its related processes are not commonly perceived as sustainability activities (Whitmarsh et al. 2011). In our survey research, while high proportions of respondents recycled newspaper and glass (93 per cent) and switched off lights in unoccupied rooms (66 per cent), few avoided eating red meat (10 per cent). Our household survey indicated that preferences for meat did change from July 2008 to July 2009, with 7 per cent increasing and 28 per cent decreasing meat consumption. However, it was not sustainability that mobilized preference changes to diets, with less than 1 per cent citing 'the environment' or 'climate change'. Instead, participants highlighted 'cost' (17 per cent), 'diet' (15 per cent) and 'health' (10 per cent). In a more detailed study of environmentally conscious households with young children, food preparation (cooking) was identified as a sustainability activity much less commonly than food production (vegetable gardens, keeping chickens) (Organo et al. 2012). Even in the most environmentally committed households, the messy realities of daily life – hungry kids and time-poor parents – dominate the process of getting the meal on the table.

A connected issue is waste. Large amounts of food are thrown away at various points in the production and consumption cycle. Australians, for example, 'throw two or three million tonnes of food waste into landfill each year' (Wright et al. 2009, 49). It is estimated that up to a quarter of all food is wasted, although reliable and consistent data are elusive (Mason et al. 2011). A further element of waste is that many people in the affluent West consume much more food than they need, carrying it around as excess body fat.

A comparison of the greenhouse gas emissions of five different Australian diets by the CSIRO (Wright et al. 2009, 52) shows that the most sustainable diet is a vegetarian one, due to the high emissions associated with meat production, particularly beef. This is consistent with a range of international studies (Garnett 2011; Berners-Lee et al. 2012).

Other simplistic concepts are associated with food sustainability. One of the most popular is that of localism, for example the concept of 'food miles' or the 100 mile diet. The concept of food miles has been extensively critiqued. Transport is only one aspect of the greenhouse gas emissions associated with food production, and there are necessary trade-offs between sustainability and international development (Saunders et al. 2006; Wallgren 2006; Mila i Canals et al. 2007). Transport accounts for only 12 per cent of food chain greenhouse gas emissions in the UK; the biggest single component is agriculture itself (Garnett 2011). Nevertheless, local food movements provide opportunities to think of food provisioning within community economies, with important ethical and social benefits (Hill 2011). Community food enterprises create spaces where 'interdependencies between people and the environment are centre stage' (Cameron and Gordon 2010, 1). Other deceptively simple concepts are organic or 'slow food', understood in opposition to the evils of 'fast food' (Petrini 2007). Critiques include inflections of class and reinforcement of representations of social distinction (Guthman 2003; Hayes-Conroy and Hayes-Conroy 2008, 467).

And of course there are many different kinds of environmental impacts associated with food production (Gerbens-Leenes et al. 2003). Greenhouse gas emissions may need to be traded off against water footprints (Page et al. 2012). Water footprints associated with food production are much larger than direct domestic water consumption. Italians average 380 litres of water consumption each per day for domestic purposes, while the estimated water footprint of an Italian margherita pizza is 1216 litres, and that of a kilogram of pasta is 1924 litres (Aldaya and Hoekstra 2010).

In this chapter we consider these and other dilemmas, illustrating the complexity and geographical variability in what constitutes food sustainability. We use the example of one popular family meal, spaghetti bolognese. Ingredients in spaghetti bolognese are variable but always include pasta and a sauce based on minced beef and tomatoes.

PASTA

Adult Australians consume over 11 kg of pasta each annually (ABS 1999), and it is also a popular children's food. The choices for pasta include mass-produced or boutique, organic or conventional, locally made or imported (usually from Italy). Our analysis of pasta draws on our wider study of wheat networks (Head et al. 2012). In this study we contrasted industrial scale manufacturing, which we called 'Big Pasta', with boutique manufacturers. Both durum and non-durum semolina flour are used at one Big Pasta manufacturer in Melbourne, the balance between them being directly influenced by the cost of transport. The premium durum wheat is grown in northern New South Wales and southern Queensland, with much higher transport costs than non-durum flour, which is sourced from southern New South Wales and Victoria. At one 'Small Pasta' manufacturer, in a country town in northern New South Wales, all the pasta the company makes is sourced from durum wheat grown on the property. They claim full traceability of their product right down to the seed used to grow the grain. In this they are typical of regional producers trying to add value to agricultural products where they are grown, rather than just exporting to city manufacturers.

There are three products at three different price points: Big Pasta's non-durum budget pasta, their premium durum product, and Small Pasta's boutique product, which is the most expensive. The latter is claimed to be a 'meal in itself', having so much flavour that there is no need for an added sauce.

What if these flours had been produced organically – would that improve their sustainability? The challenges of organic broad acre cereal farming are different to those in say a permaculture vegetable garden. Organic agriculture operates on longer fallow phases than conventional cropping (Derrick and Dumaresq 1999). Organic farmers achieve lower yields, but command a price premium five or six times higher than conventional wheat (Head et al. 2012, 98–9). Organic farmer Arthur explained his methods of weed removal to our researchers, who reported:

> Where weeds are present in his crop they must be removed without the use of chemicals, usually by physical removal using disk ploughs which were described as an 'old fashioned' farming practice, similar to how people farmed before herbicides were available. The soil is cultivated using the disk plough. (Head et al. 2012, 99)

A major drawback of this system of repeated physical cultivation is disturbance of the soil profile, and increased rates of wind and water erosion. In conventional Australian wheat farming, the adoption of no-till or conservation-till systems since the 1980s has greatly reduced soil erosion; nevertheless: 'these technologies are expensive and size efficiencies are required'

(Head et al. 2012, 99). This was not an option for Arthur, even with the premium prices available for his organic crops. The relatively small scale of organic wheat farming makes for other sustainability contradictions; a miller focusing on producing organic stone ground flour in northern New South Wales has to source grain from across the continent in order to ensure consistency of supply.

What about the water footprint of wheat crops? As wheat exports around the world are predominantly from rain-fed agriculture (Australia providing a prime example), international trade is considered to be contributing to global average water savings: 'import of wheat and wheat products by Algeria, Iran, Morocco and Venezuela from Canada, France, the US and Australia resulted in the largest global water savings' (Mekonnen and Hoekstra 2010, 1265). Or, put another way, 'without trade the global wheat-related water footprint would be 6 per cent higher than under current conditions' (Mekonnen and Hoekstra 2010, 1273). Most Australians would be surprised to hear that their wheat exports contribute to overall global water savings. However 'the costs of water consumption and pollution are not yet properly factored into the price of traded wheat, so that export countries bear the cost related to wheat consumption in the importing countries' (Mekonnen and Hoekstra 2010, 1273). If an Australian household is eating pasta imported from Italy, their biggest environmental impact is not the transportation costs, as they might have thought, but the importation of virtual water from Syria and Iraq, key source countries for Italian wheat imports.

If there is bread as well as pasta on this family table, the sustainability hotspots of its lifecycle are the electricity use in retail and consumption, and in storage and processing, and the insecticide and fertilizer use in crop production (Narayanaswamy et al. 2004). The low proportion of impacts associated with transportation challenges the ideal of local food as a solution. While farmers often get the blame for the environmental impacts associated with food production, on this evidence that is only partly justified. Certainly changes to on-farm practices could reduce the detrimental effects on terrestrial and aquatic ecosystems of fertilizer and pesticide use. But these contribute much less to global warming than do the storage and processing phases of flour milling, and subsequent retail and consumption (Narayanaswamy et al. 2004).

Perhaps the biggest issue with the sustainability of grain production is the significant proportion of the crop that is used as stock feed, as we move higher up the food chain. That brings us to the next key ingredient in spaghetti bolognese.

MEAT

Greenhouse gas emissions associated with beef 'will vary greatly depending on how it is produced, how and where the cattle are raised, and how much land clearing or re-clearing is carried out' (Wright et al. 2009, 52). In a US study of greenhouse gas emissions associated with beef production, 'enteric methane is the leading factor, although both feed production and manure management (primarily nitrous oxide emissions) also make substantial contributions' (Pelletier et al. 2010, 383). But this varies across different production strategies. Pelletier et al. found that pasture-finished beef

> from managed grazing systems as currently practiced in the US Upper Midwest is more greenhouse gas intensive than feedlot-finished beef... when viewed on an equal live-weight production basis. This conclusion is consistent with previous research, which has shown that higher quality diets and increased growth rates reduce ruminant methane and manure nitrous oxide emissions, both of which are key contributors to life cycle emissions. (Pelletier et al. 2010, 386)

While this finding is somewhat surprising, they note several important distinctions between pasture-finishing in this region and unmanaged rangelands (such as those that characterize much Australian beef pastures). Grass-feeding in the Upper Midwest includes substantial proportions of hay, particularly in winter. Hay production and transportation has its own energy costs. Further, these managed pastures require extensive inputs in the form of fertilizer and seeding, again in contrast to unmanaged rangelands. They further emphasize that 'none of the systems analysed can be described as ecologically efficient relative to most other food production strategies' (Pelletier et al. 2010, 388).

One of the ways in which Australians have considered addressing this dilemma is through eating kangaroo. In Australia, 11 per cent of national greenhouse gas emissions have been attributed to the livestock industry (National Greenhouse Gas Inventory 2009). Diesendorf (2007) called for a 20 per cent reduction in national beef consumption, suggesting the option of eating kangaroo. Here the logic is that kangaroos produce less soil erosion (without hooves), and less methane than cattle because of their different digestive tract (Morrison et al. 2007). Wilson and Edwards (2008) modelled the removal of 7 million cattle and 36 million sheep, and increasing the kangaroo population to 175 million, and found that national greenhouse gas emissions would be reduced by 3 per cent by 2020. The *Garnaut Review* (Australia's equivalent to the *Stern Review* in the UK) underscored the significance of eating kangaroo as a more sustainable option than conventional livestock (Garnaut 2008). Environmental scientists also appealed to the logic of eating kangaroo as a more sustainable red meat, given the contribution of cattle and sheep to land degradation in the semi-arid sheep rangelands (Grigg 1995).

All Australian states legalized the commercial sale of kangaroo meat by 1993. The National Food Authority classified kangaroo flesh as a 'game meat product' in 1995 and various marketing campaigns ensued. Since 2000, kangaroo sausages, fillets and roasts have become more widely available, including in many supermarkets. Yet kangaroo is not part of most Australians' weekly practices of food shopping, cooking and eating. A national survey in 2007 suggested that only 4.7 per cent of Australians eat kangaroo monthly or more frequently (Ampt and Owen 2008). Consistent with these government statistics, respondents in our household survey reported consuming chicken (90 per cent), beef (85 per cent) and lamb (52 per cent) each week, but only 8 per cent ate kangaroo weekly.

Why hasn't kangaroo meat become a normal ingredient in weekly diets? In our household ethnography kangaroo meat was 'bush', 'specialty', 'restaurant' or 'exotic' food. Consistent with earlier research (Purtell and Associates 1997), kangaroo meat, when eaten, was a meal reserved for special occasions. Other people said they would never eat kangaroo, and thought of it as inedible. Some people said kangaroo meat should be used only for pet food given many farmers see kangaroos as a pest. Others treasured the kangaroo as a cute animal or as a symbol of Australian national identity. When asked if there were any foods they would never eat, Alan and Sam responded:

> Oh no, what I wouldn't eat, for start, a bloody emu or kangaroo... just wouldn't eat kangaroo or emu just on principle. The bloody national anthem, national icon... Why would you eat them? I think it's wrong. Americans wouldn't saddle up to a bloody big golden eagle would they? (Alan, retiree, aged 62)

> I don't know, there is something sort of strangely weird about eating your national symbol isn't there? (Sam, student, aged 46)

Such opinions have historical, colonial origins: as Craw (2008, 90) argues, the kangaroo was elected as an Australian national symbol against a long-standing background of European scientific 'fascination with the "otherness" of Australian nature'. But in so doing, British colonizers locked kangaroo meat out of cookbooks, practices and habits, requiring the establishment of an unsustainable livestock industry based on cattle and sheep.

Just as getting the family meal on the table involves many considerations other than sustainability ones, so the eating of kangaroo involves an additional complex of factors beyond our discussion here, including ethics (Probyn 2010), indigenous land management (Davies 1999), and difficulty of farming (Grigg 1987).

TOMATOES

The tomatoes used in spaghetti bolognese are typically canned or pureed, often with the addition of tomato paste. Tomatoes are one of the most commonly grown things in home vegetable gardens so, at the right season, some households may be using these. We found that backyard food production must be understood as producing a variety of social goods, echoing Gaynor's interpretation that 'where food is produced by a household, it is often produced not as "food", but as "home-grown food", a distinctive category of produce' (Gaynor 2001, 27). These social goods include reproduction of tradition, education of children and connection to the soil, as well as food valued for its freshness and taste (see also Chapter 16). For example, retired Wollongong resident and Macedonian migrant Slave's explanation was clearly about more than calories:

> Every tradition from there [Macedonia] we brought it here, so that's why we are making gardens and trying to produce chillies and tomatoes or something little, in a way to remind us of there, but somehow to have the taste from there brought here, and to continue the tradition of making produce and something from the soil. (quoted in Head and Muir 2007a, 99)

The considerable labour involved in serious home production invokes diverse responses. Working mother Karen somehow found the time to produce substantial quantities, explaining that, 'I can be almost self-sufficient in the summer time growing lettuces, and tomatoes and beetroot' (Head and Muir 2007a, 92). On the other hand, people commonly talked about wanting to grow food – invoking ideas of getting back to nature and becoming connected to the soil – rather than actually doing it. Lynette described her plans for the future thus:

> I'd like to get back to nature and be able to put a spade in the garden and to get the satisfaction of seeing something grow, something productive and something that you can eat, instead of just going to a supermarket and buying everything in a plastic bag, and hopefully we won't be using pesticides. (quoted in Head and Muir 2007a, 93)

So it is most likely that the tomatoes in our spaghetti bolognese have been bought from a shop. As there is more information available on the sustainability issues associated with fresh tomatoes, we use those here as the main example. (Note, however, that as Pritchard and Burch (2003) argue, the fresh and canned tomato industries are almost completely separate.) In this section we draw particularly on Roggeveen's (2010, 2012) study of tomatoes sold in Sydney's wholesale fruit and vegetable markets. Roggeveen combined life-cycle analyses and ecological footprint studies with a relational, qualitative 'following' approach (cf. Cook 2006). The latter uses ethnographic methods to pay attention to the human and non-human complexities of food and other commodity chains. Roggeveen found that on-farm greenhouse gas emissions – mostly associated with greenhouse heating (the dominant mode of production) – were far greater than emissions from fuel used for transport to market, per unit of tomatoes. They also appeared much greater than greenhouse gas emissions by wholesalers or retailers. Tomatoes sourced locally are not necessarily more greenhouse gas efficient, nor are those from small or lower-tech farms.

These findings are consistent with Page et al. (2012) who compared field tomatoes with low, medium and high-tech greenhouse production. They found that 'Hotspot areas for reducing carbon emissions are identified to be the cultivation stage for all the greenhouse production systems (contributing 54–82 percent carbon emissions within cradle-to-plate)' (Page et al. 2012, 223). Coal and natural gas used in horticultural greenhouse heating was identified as a major issue in both studies.

Page et al. also demonstrated that there are necessary trade-offs between water and carbon footprints: 'For instance the low-tech greenhouse production had the lowest carbon footprint while the field production had the lowest water footprint' (Page et al. 2012, 223). The high-tech greenhouse had the highest carbon footprint but a very low water footprint.

Part of Roggeveen's initial aim was to assess the potential for more detailed labelling at the consumer end. She argued that the appearance of a smooth logistical flow maintaining supply to shops is somewhat illusory. In contrast,

> the production-supply chain has elements of friction and competition as businesses negotiate the daily weather and economic conditions of selling a fresh commodity. There are many combinations of the potential tomato journey, and they can change daily, as actors buy and sell their tomatoes from and to different people and use different transport options. Businesses throughout the supply chain aim to control as much as possible the weather's impact on their goods, and other business risks such as breaks in energy supply. (Roggeveen 2012, 22)

Roggeveen therefore concluded that under present circumstances industry-wide carbon labelling is unlikely to be a practical tool to reduce greenhouse gas emissions from fresh fruit and vegetable supply chains: 'the integrity of the labelling would be reliant on many actors in the chain undertaking accurate and considerable data measurement and labelling, possibly on a daily basis' (Roggeveen 2012, 22). The difficulty of providing comparative labels on different types of tomato, perhaps sitting side-by-side in a shop, is magnified when we consider what we understand as the total agri-food system. The pathways from paddock to shop, and then to plate, are many, and more accurately described as a dense web or network of processes.

WHAT TO DO?

Tomato webs intersect with wheat webs intersect with meat webs. Even different meals of spaghetti bolognese will have different patterns of environmental impacts. And of course it is more complex than that. We have left out the topping – freshly grated parmesan cheese and other important ingredients such as garlic and basil. Nor do we have space here to consider the accompanying salad or dessert, or the possibility of eating at an Italian restaurant rather than at home. These food webs do scale up in ways that have significant implications for the sustainable management of landscapes, that aggregate to produce the greenhouse gas emissions of a monolith summarized as 'food', and that create more or less food security for different parts of the population.

There is no end to the potential for detailed anxieties around food, and any trip to the supermarket involves hundreds of individual decisions as it is. Households seeking to contribute to more sustainable outcomes can probably be more constructive by attending to general patterns than by getting too hung up on single foods. There are huge issues of consistent data measurement and, at the moment, communicating with the consumer, for example through carbon labelling, is highly fraught. Likewise single issues, such as local food, or food miles, will almost always be too simplistic to provide total solutions. This is not to argue that we should be flying 'exotic foods' around the world, rather that in a system in which self-provisioning is not possible for the current population levels, and their urban concentration, some global trade is likely to remain part of the total package. Nevertheless community food initiatives do catalyse grounded engagements between humans, plants, soil and water, and encourage ethical thinking including (and beyond) carbon emissions impacts. Other macro-scale systems might not seem more efficient now, but that does not mean that localized experiments are pointless, or that they could not scale up in effective future ways – especially as circumstances change.

Low greenhouse gas food behaviours are broadly consistent with a diet of fresh, simply prepared food (Wright et al. 2009), though this is deceptively simple. Gendered responsibilities within homes place the burden of cooking and preparing food on women; this combined with longer average working hours and the drift towards multiple part-time jobs and shiftwork hamper abilities to cook daily fresh meals. Fresh fruit and vegetables are not cheap, nor necessarily accessible – even in modern cities (Bakar 2010). Pressed for time, and with limited means, families make all kinds of food decisions, many imperfect, that sit outside the healthy eating mantra. Meals are a key point of ethical juggling within households, and are enmeshed in wider rhythms of work and city life.

As will also become clearer with other themes in this book, the burden of acting sustainably in relation to food ought not to be lumped solely on consumers. This is a shared responsibility. The complex webs of production and supply also offer many different points of broader societal intervention, and different pathways of policy traction.

3. Clothes

For all but the most committed nudists, clothing is an essential need. It is thus a central, though far from straightforward, part of the household sustainability story. Clothing accounts for between 5 and 26 per cent of total household water use, up to 14 per cent of total household waste, and between 7 and 10 per cent of total ecological footprint, according to methodology and country (Tukker et al. 2006; ACF 2007; Kenway et al. 2008). The purchasing, use and disposal of clothing generates on average 0.8 tonnes of CO_2-e per household per annum (or 8000 garbage bags filled with CO_2) (Wright et al. 2009). Clothing also contributes to other ecological problems such as excess water use, and groundwater, soil and air pollution in its production, distribution and consumption.

What further complicates the dilemmas of sustainability surrounding clothing is that, as an industry and practice of consumption, clothing is also fashion: most people own more clothing than they need, and only regularly wear a fraction of it, replacing and replicating items seasonally (or even more often) without necessarily needing new clothes (Gibson and Stanes 2010). Clothing is, arguably, the exemplar commodity to illustrate the difference between an object's use value (its basic utility to humans) and its exchange value (what that object is worth on the market based on what people are prepared to pay). Our behaviour with regards to clothing is only partly based on its use value; instead we buy, wear, dispose and replace clothing based on aesthetics, appearances, and emotion. This difference is what distinguishes clothing as a *cultural* industry – whereby the cultural logics of fashion, subculture, marketing and identity deeply shape the organization of its manufacture, and consumption practices within the household. Clothing is practical, ubiquitous and necessary, but also indulgent, often impractical, decorative and symbolic.

People have always used clothing to stay warm or cool. Clothing has long helped us to be modest or to attract suitors, to construct a sense of ourselves, to conform to social norms (or to stand out from a crowd), or to judge others by appearances. Clothing has always been linked to status and fashion-consciousness: in ancient Egypt, in China, in the Aztec Empire (Ross 2008). In Europe in the 1600s rare materials, fine sartorial skills and fashion styles (from the number of ruffles to exotic dyes) marked the social classes, signalled elevation from manual labour and maintained class distinctions (Pears 2006).

Mass production technologies democratized clothing in the twentieth century, but also created the prospects of stockpiling clothing beyond immediate needs. Generations came to be defined by particular styles – the flared jeans and platform shoes of the 1970s; the fluorescent colours of the 1980s; punk's safety-pin aesthetic – and from New York to Mumbai clothing now conveys a sense of identity, of being 'up' with the latest trends, bestowing popularity, caché and a sense of sophistication. For these reasons and many more, people invest time and money in clothing – accumulating too much clothing that then requires larger wardrobes and additional storage space (or divestment of 'old' items that might be perfectly usable, but that weren't favourites, or didn't quite fit right). In extreme cases an addiction to buying clothes is the prime reason for escalating, and unmanageable credit card debts – an illustration of the social and emotional power of fashion (Pears 2006). Consequently clothing consumption is now deeply embedded in a host of cultural, social and psychological processes. Disentangling these in the context of household sustainability is an enormous challenge.

SUSTAINABILITY ISSUES

Upstream sustainability issues for clothing include raw material acquisition and the nutrient, energy and water use required to produce wool, cotton and other natural fibres (as well as loss of habitat converted to grazing land for sheep or to cotton plantations); energy, water and chemical inputs required for synthetic fibre production (as well as pollutant by-products from the use of petroleum in creating polyesters and nylons); and fuel use in transporting raw materials to processors, and from processors to fabric wholesalers, manufacturers and retailers (and from shops to the home). To illustrate, the average suit generates approximately 300 kg CO_2-e (or 3000 garbage bags) in its production (Wright et al. 2009).

Once in the home, sustainability issues for clothing include water and energy used by people looking after and cleaning clothes; impacts of the use of phosphate-rich detergents on water and soil systems; how clothing is disposed, reused or recycled; and the sheer quantity of clothing purchased in relation to that which is essential to stay cool, warm, comfortable and to maintain appearances. What is not clear to consumers is exactly how significant are the upstream impacts of a given piece of clothing: they can vary substantially between fibre type and even within fibre types (Table 3.1). Cotton, for example, once considered more precious and valuable than silk, requires enormous amounts of water (between 7000 and 29 000 litres per kg produced, varying markedly with country and region of production), while (raw) wool requires only 125 litres per kg (Fletcher 2008). Wool also uses only about a fifth of the

Table 3.1 Measurable impacts of fibre growth and production

Fibre	Water Consumption (per Kg)	Energy Consumption (per Kg)	Environmental Impacts
Cotton	7000 – 29 000 L/kg	48.65 MJ/kg	Use of agrochemicals (fertilizers and pesticides) and soil contamination Water consumption and contamination Soil erosion and degradation Processing Land displacement
Wool	125L/kg Raw wool	8 MJ/kg	Overgrazing (soil erosion and degradation) Pesticides
Acrylic	–	157 MJ/kg	Non bio-degradable Porous fibre which holds chemicals within it Uses petrochemicals
Polyester	–	109.41 MJ/kg	Energy intensive Uses petrochemicals Non bio-degradable fibre Hazardous chemical emissions
Nylon	–	150MG/kg	Energy intensive Hazardous air-borne emissions Contamination of water Non bio-degradable fibre Uses petrochemicals

Source: adapted from Stanes 2008.

energy required to produce cotton. Yet whereas woollen tweeds, knits and gabardines once dominated garment construction, since the 1980s worldwide textile production has shifted substantially to cotton (now accounting for 40 per cent of all textile production) reflecting consumer preferences for the feel and climate comfort of cotton (although fine merino wool is fighting back). This has come at a considerable ecological price. Nevertheless, both cotton and wool entail substantial alterations of ecosystem and landscape, with wool production in Australia (long a world leading producer and exporter) responsible for dramatic landscape changes and problems with habitat loss and soil erosion.

Synthetic fibres minimize landscape impacts, yet are much more energy intensive (more than 15 times that of wool, in the case of nylon), and depend on the petrochemical industry (Table 3.1). A range of alternative and experimental 'eco' fabrics have hit the market, which reduce water and energy use, and recycle existing materials. These include organic cotton, fabrics made out of recycled water bottles (that feature in some brands of surfwear and outdoor hiking gear), and a revival in the use of hemp. But all these still require inputs, have distribution impacts and need washing, storing and mending after purchase. The potential for greenwash in fashion is ever-present, given the dependence of the industry on consumers stockpiling clothing, continual seasonal (and sub-seasonal) purchasing, and limitless consumption. Nowhere have manufacturers promoted sustainability by recommending reducing or resisting buying more clothes and, as Wright et al. (2009) vividly put it, every dollar spent on new clothing produces on average six garbage bags of CO_2-e.

Fast-Fashion

The recent trend towards *fast-fashion* has exacerbated ecological impacts and the accumulation of clothing. A parallel to fast food, it is the making, selling and consuming of large quantities of lower-quality fashion at an increased velocity, at lower prices (Doeringer and Crean 2006). Fast-fashion represents a qualitative shift in the system of provision for clothing: it links tactics by manufacturers to increase throughput with those of retailers to encourage more frequent visitation by consumers (see also Chapter 8). The result is increased impulse and out-of-season purchases, which might be good for manufacturers and retailers, but consequently ramps up sustainability impacts. Fast-fashion is premised on the rapid cycling of fashions available in shops, on even a fortnightly basis, and a lower price point for consumers (who would not be able to sustain ongoing repeat purchasing at elevated price levels). Capitalizing on the desire of people to look good in the 'latest' clothes, this is a deliberate tactic to speed up the fashion cycle beyond the usual annual seasonal rhythms. What makes this possible is that fashion is highly price elastic: if prices fall considerably many people feel encouraged to buy more – whether or not that clothing is needed (Gibson and Stanes 2010). Allwood et al. (2005) analysed the demand for low cost disposable clothing in the United Kingdom and found that over four years, the number of garments purchased per person had increased by over one third.

Accordingly the mix of priorities in the design and manufacture of fast-fashion has shifted. Quality of fabrics can be reduced because fast-fashion is premised on rapid cycling of clothing in shops, and through the consumer's wardrobe. The logic is that if fast-fashion wears out quickly with repeated washing, if it fades, stretches or tears (which it does, generally, because it is poorer quality) this matters less, because it will be quickly replaced – or have already been replaced – by other items bought at low prices. Likewise labour costs are minimized in fast-fashion through the exploitative use of sweatshop work, outsourced factory production in low labour-cost countries, and piece-work within industrialized nations (frequently migrants working at home). A higher proportion of the costs of production are invested in design, with the model premised on streetwise looks and the 'thrill of the new'. Upstream sustainability impacts include: higher overall volume of production of cloth-ing, with commensurate elevated levels of raw materials, water, energy, fertil-izers and chemicals used in production; higher transport-related impacts (through higher volume of clothing made and traded); intensification of fertil-izer and pesticide use in fibre production (in order to reduce costs of raw mate-rials) and increased exploitation of workers. Further confusing the issue is that some high price-point brands now also use fast-fashion production methods and marketing tactics. Difference in price is attributable instead to the caché of brand name. Either way, poorer quality clothing, and more of it, means an exacerbation of energy, water and ecological problems in the production and consumption of clothing.

Laundry Practices

Beyond fast-fashion, a significant proportion of fashion's overall environmen-tal impact is post-purchase (see also Chapter 7). Washing and drying consumes up to 85 per cent of the total energy for the lifecycle of clothing, while heating the water for washing machines is responsible for 50 per cent of all energy consumed (Richter 2004; Stanes 2008). Meanwhile the transporta-tion of a conventional cotton shirt only accounts for 8 per cent of its total envi-ronmental cost (Allwood et al. 2005). Dryers are pernicious, consuming on average 5.7 kW per hour – but their use varies nationally and with household type and structure. In colder climates their use is far more pervasive. In Australia, where line-drying is still the norm, 50 per cent of the households that use dryers live in flats, apartments or units, an increasingly common housing type where strata laws frequently prevent tenants from visibly drying clothes on balconies (Stanes 2008).

Washing practices too have changed over time. Washing clothes was once labour-intensive and complicated; worn clothes were aired before re-wearing and some measure of odour and soiling was deemed 'normal' (Chapter 7). Only the wealthiest could afford paid staff to regularly wash clothes and maintain a pristine appearance. Accordingly, fabric types (especially woollens) were used that aired effectively and that could be quickly scrubbed into an acceptable state for re-wear. The invention of the automatic electric washing machine in 1937 and subsequent uptake from the 1950s to the 1970s revolutionized laundering practices in the West, altering our relationship with clothes, alongside norms of bodily cleanliness and smell (Shove 2003). Research from the Netherlands has showed that the average piece of clothing stays in a person's wardrobe for three years and five months, is on the body for a total of 44 days, and during this time is worn on average between 2.4 and 3.1 days between washing (Uitdenbogerd et al. 1998), but this varies enormously with items, and cultural preference. Washing clothes after only one wear, even when only marginally 'dirty', is normal in some countries, and any sign of odour or soiling can be socially unacceptable, especially in white collar work where the 'corporate look' often demands impeccable appearances (although men's and women's suits continue to be re-worn many times between visits to the dry-cleaners) (Pink 2005). Subtle shifts have meant that, for instance, cleanliness is viewed as 'freshness' and 'whiteness' rather than being germ-free, related to the widespread marketing and use of detergents in washing machines. The sum effect is that clothes are now washed many more times and hence wear out more quickly than they once did, which exacerbates overall resource use and carbon emission impacts, and indirectly (through wearing-out of clothing) heightens dependence on the fast-fashion system.

In our household ethnographic work, pre-World War II norms of cleanliness lingered within one household (in contrast to Shove's (2003) argument about disinfection losing ground to freshness as the rationale for laundering). Jan and Josefa described themselves as environmentally conscious, yet still debated the use of hot washes for their presumed capacity to kill germs:

> Josefa: When I'm washing the bed linen for our bed and my clothes and undies and stuff I have it on hot and warm.
> Jan: I have read recently, there's a research article that was reported in the paper that said the temperature makes no difference to the cleanliness of your clothes, so if you wash them on cold it's just as good.
> Josefa: I'm not talking about cleanliness, I'm talking about getting rid of bugs and things.

Jan: Yeah, no, that's what it was saying, cleanliness meaning getting the bugs and things, that in fact, apparently cold is no different from hot and that it's a furphy [ed. note: Australian slang for a rumour or improbable fact] about washing it in very hot water.

Josefa: Well that's not what the asthma people told me.

Jan: Oh. Oh well, anyway that's science for you.

Such everyday debates illuminate the extent to which dilemmas of clothing and sustainability are wrangled within households.

Waste

Additional impacts arise from clothing waste practices: the United States Environmental Protection Agency Office of Solid Waste estimated that Americans throw away 30 kg of clothing and textiles per person per year (Claudio 2007). In the UK the amount of clothing and textile waste created annually is nearly 40 kg per person (Allwood et al. 2005). Barely a quarter of this waste is reclaimed; half of which is used by material recovery firms, and the other half incinerated. The other three-quarters of waste textiles are sent to landfill (Fletcher 2008), where textile waste is responsible for toxic chemical leaching and groundwater pollution. Hamilton et al. (2005) estimate that approximately AU$1.7 billion (AU$1700 million) is spent annually in Australia on clothing that is never worn.

TRANSFORMATIONS AND COMPLEXITIES

As knowledge spreads of the water and energy use impacts of warm washing and using electric clothes dryers, government initiatives have encouraged manufacturers to recommend cold washes and line drying on care labels (a child's organic cotton t-shirt bought by one author recently had this on its washing instructions tag: 'save the climate – wash cool – line dry'), and at least one global jeans brand now labels their items to be aired rather than washed for at least six months to lengthen product life, reduce water and energy use and to enhance the 'ageing' of the denim.

Such efforts are, however, counteracted by other structural impediments: some master-planned estates in the United States ban drying of clothing outside on lines, compelling residents to use dryers. Communities have launched campaigns aimed at state and federal legislators to introduce 'Right to Dry' legislation – seeking to overturn clothes-line bans (Rosenthal 2008). Meanwhile the prevalence of air conditioning (in tropical and warm temperate climates) and central heating (in cool climates), in both the home and at work, has shifted the market for and availability of clothing suited to a range

of extremes (see also Chapter 5). In Australia for example it is extremely difficult to buy men's and women's suits for hot summers. Whereas once linen
fabrics and the 'safari suit' (that icon of 1960s fashion) met demand for working
in hot summer offices, now an 'all-season' wool-blend suit is the norm, available
year-round but effective only in temperature-controlled environments. Here a
resource-intensive form of climate control and the structure of the fashion industry work in tandem to block improvements in sustainability (Chapter 5).

Other efforts include the shift towards second-hand clothing consumption,
from charity shops to dedicated commercial vintage stores. Second-hand clothing consumption most obviously reduces the overall production of clothing so
long as each second-hand item purchased negates the need to purchase another
item new. The difference that buying second-hand clothing can make is huge:
in Australia, the Salvation Army recycles 2.8 million pieces of clothing annually through its 105 charity shops (Wright et al. 2009, 45). Second-hand clothing enables people to buy 'classic' or 'retro' items from an era when clothes
were better made to last (although from the 1960s through to the 1990s it is
synthetic fibres that have lasted better than cottons and wools). Second-hand
clothing offers the fashion-conscious consumer a sense of distinctiveness, and
the thrill of finding a unique item no longer available en masse (Pears 2006).
Yet vintage clothing has itself become a high throughput industry, a business
model that 'begins to resemble fast fashion' (Siegle 2013, 1). Other reactions
are the 'slow-fashion' movement (which emphasizes 'classic' designs and well-
made items that are built to last); and the revival of hand-made clothing, made-
to-measure and bespoke tailoring. The latter is still a province of the rich, but
certainly premised on reduced overall volume of clothing production and
reduced waste, through more regular wear of clothing that fits perfectly.

Notwithstanding such efforts, it remains the case that people consume
clothing differently, with sustainability implications. Not everyone is comfortable wearing second-hand clothes (especially undergarments closer to the
body – Gregson 2007), or has the time to get bespoke items made to measure,
and the hefty price tags on high quality items are off-putting to most. Clothing
consumption is clearly shaped by underlying values and psychology as much
as any environmental awareness: so that those inclined towards thrift and
making-do, as opposed to excessiveness or abundance, are more likely to have
smaller wardrobes, fewer clothes, and are more likely to mend, sew and be
satisfied with slightly worn or dated items. For others, clothes are deeply
linked to a sense of self-worth and identity, and linked to appearances, so that
'shopping addicts' and 'fashion victims' may also use clothes as a means to
improve self-confidence, or express elements of their personal identity.

There is also a generational dimension: Generation Y are said to have
different perspectives on clothing consumption from those who grew up
through the depression and war years (or who had parents who did) (Stanes

2008). Perhaps uncertainty surrounding one's identity or place in the world is a ubiquitous feature of youth that generally fades with age: hence much fashion marketing is aimed at the young, who may be more likely to stockpile clothing, and least likely to have a mortgage, a baby or other responsibilities.

Many are simply constrained by circumstances, with frugality concomitant with necessity. One of our ethnographic participants, Paul, lived in public housing and relied solely on the aged pension for his income. He felt a strong sense of stewardship of nature and an aversion to waste:

> I'm a child of the depression of the 1930s. I was born just after the depression but my parents, that was their style of living and that was what I picked up and my siblings picked up as values. They were all alive at that time. And I just have a feeling that you use things up ... 'til they haven't got much more value and then you throw them out. Or use them for cleaning cloths or whatever.

His wardrobe was sourced entirely from charity shops – except for socks:

> The only thing that I haven't bought from the [charity shop is] the socks because I bought those at the el cheapo shop down town the other day, but everything else I'm wearing is from the op-shops. At my age you don't have to worry about being sort of, you know, the man about town, stylishly dressed, it doesn't matter, so long as something works. As an old aged pensioner I just feel I could buy expensive stuff but what the hell's the use of that? It's not worth it. And I also have ... got a big feeling for using the resources that are around about us.

For others on paltry incomes, the low prices of fast-fashion (in some cases *scandalously* low) at least makes clothing accessible – even if poorly made and with a high upstream environmental cost. The poorest households were the least likely to own clothes dryers. There are paradoxes too at the other end of the income scale: those with high disposable incomes have the capacity to accumulate it and to have large houses with ample storage space, but they are also more capable of buying high quality, high price items that last.

Such tensions are everywhere. In Elyse Stanes' ethnography of clothing consumption (Gibson and Stanes 2010) there were contradictions within types of consumers: 'fashion victims' were more likely to own far more clothing than they needed, but were more likely to be interested in the environmental and social justice impacts of clothing production. Frequent consumers of fashion, usually women, were more likely to buy 'eco' fabrics, to purchase second-hand clothing and to share clothing with friends and family. They were more likely to look after their clothing with an eye to extending its life, and mended and sewed regularly. By contrast, those who were resistant to clothing as fashion, and instead prioritized utility (usually men), were least likely to be aware of the environmental and social justice implications of clothing production, were least likely to sew and mend, least likely to pay more for longer-lasting 'classic' items, and were most likely to buy poorly made items at low cost that would need replacing more often. Yet they owned the fewest clothes in total, eschewing stockpiling and delaying the task of shopping for new clothes as long as was humanly possible.

WHAT TO DO?

The above dilemmas of clothing consumption and sustainability are difficult to resolve. In the fleeting moments of shopping, upstream ecological impacts are simply not apparent; without accurate and detailed labelling (not yet pervasive for clothing in a manner like fridges, Chapter 12), it is impossible to know which of two competing brands of jeans has the least water and energy use impacts. It is worth exploring fabrics that absorb less sweat (and therefore require less washing) or are made from fabrics with demonstrably less resource impacts (such as hemp). The truism nevertheless remains that the single most significant improvement one can make is through buying less clothing.

Failing this, quality clothing lasts. Clothes can be washed less often (Chapter 7), and in low temperatures. Dry on a line, and choose styles that are 'classic' and therefore less likely to date or appear embarrassing over time. We could all share more clothing with others and attend (or host) clothes-swap parties. It is naïve to assume that people will abandon clothing as fashion. Clothing will remain a part of self-expression and identity formation, because humans are social, judgemental, emotional beings. And thus a tightly frugal approach to clothing is likely to be exceptional, unless macroeconomic circumstances deteriorate substantially. Second-hand clothes provide possibilities to minimize new production and expenditure on clothes and still express a preference or taste – even to build a unique 'style'. Related issues persist beyond the immediate scope of the household: better labelling of clothing to reveal total environmental costs and labour conditions in production; how to 'break' the fast-fashion cycle; removing regulations that prevent line-drying clothes; improving high school curricula regarding sewing and mending clothing; better design of homes and workplaces that stay warm or cool without artificial climate control (Chapter 5); and wider availability of better-designed clothes for climate extremes. Presuming that nudism is unlikely to catch on as a widespread practice, as a species we will continue to consume clothing, and it will thus continue to present some of the most entangled sustainability dilemmas for households.

4. Water

The combination of climate change, population growth and increasing affluence is putting pressure on water resources in both urban and rural contexts. Climate change will have most of its social impact through water-related phenomena (Falkenmark 2008). Water footprints (Hoekstra and Chapagain 2007) are now often calculated alongside carbon footprints. Water supply is clearly a global issue, but it is a regional resource requiring varied policy solutions. Unprecedented recent droughts in southeastern Australia and the southwestern United States have brought water issues strongly into public consciousness, and provide insights into how we might adapt better to such conditions in the future. At the same time, it is important to remember that some 1.1 billion people worldwide still lack access to an improved water supply, and up to 2 billion experience regular water scarcity (MEA 2005). The examples in this chapter are mainly from Australia, the driest inhabited continent, in which the challenges of sustainable water supply for affluent populations are increasingly felt. This context exemplifies issues of wider international relevance.

As one water manager says, 'we must never take a water-only view' (Kelly 2004, 38), because some proposed solutions to save water in fact create larger environmental footprints through the energy use and infrastructure required to move it around, clean it up and pump it to homes. At the macro-scale, the water industry is energy-intensive – although in geographically variable ways. It contributes around 1 per cent of national greenhouse gas emissions in the United Kingdom (Water UK 2009, 1). For one shire council in rural New South Wales (NSW), 48 per cent of its greenhouse gas emissions came from water and wastewater, primarily through pumping stations, sewage treatment plants and dam aeration systems (Still 2008). At the household scale, Wright et al. (2009, 14) estimate that supplying water to a typical three-person Australian household produces 270 kg of CO_2-e, about half of one per cent of the total greenhouse gas emissions for such a household (Wright et al. 2009, 4). Households are implicated in the sustainability issues of the water they consume both directly (around 70 per cent of the water in greater Sydney, for example) and indirectly (about 70 per cent of Australia's total available water is used on irrigation for food production). These proportions are paralleled elsewhere; 70 per cent of worldwide water use is for agriculture (MEA 2005, 2).

This chapter examines the role and responsibility of the household in driving or maintaining change, given an underlying infrastructure developed in the twentieth century around what Sofoulis (2005) has dubbed 'Big Water' – the fantasy of endless supply. To what extent can behavioural and technical changes translate into real and sustained water savings, given the infrastructure of supply? Uptake of innovative domestic solutions such as dual-flush toilets, low-flow shower heads, rainwater tanks and front-loading washing machines has expanded dramatically in the last few years (Sustainability Victoria 2011). However, we will show, using the example of rainwater tanks, that technical changes alone do not necessarily result in the expected water savings, because there is a complex interplay between the technologies and infrastructure, everyday practices and cultural perspectives.

Davison (2008) traced the relationship between per capita water consumption and the infrastructure of supply in Australia's major cities. Average daily consumption has nearly trebled since the nineteenth century, with the installation of piped water and underground sewerage. Health and sanitary concerns rather than concern over water scarcity drove these developments, as they did across the global north. Twentieth century approaches continued to deal with the supply side, using engineered solutions (Kaika 2005, Sofoulis 2005, Gleick 2010), but it is widely recognized that this cannot continue indefinitely. The amount of water impounded behind dams quadrupled since 1960, and three to six times as much water is held in reservoirs as in natural rivers (MEA 2005, 2). Water sustainability needs to be managed by decreasing demand as well as increasing supply (Macdonald 2010).

More recent policy responses have focused on behavioural changes as well as technological solutions such as desalination plants, recycling and new dams. However, the dominant management model of behavioural change is rather linear in its approach, assuming that if you change attitudes you can change behaviours, and that people have a high degree of choice in these things (Sofoulis 2010). The research we draw on here has a focus on everyday practice, or 'inconspicuous consumption' (Allon and Sofoulis 2006), indicating a much more complex set of pathways and configurations.

TRANSFORMATIONS AND COMPLEXITIES

Policy makers have recognized that 'a key to improving efficiency is understanding where, when and why we use water' (Gleick 2010, 21302). The so-called Millennium Drought in southeastern Australia in the first decade of the new millennium threw light on and de-routinized many household water practices. A body of cultural research on urban water practices now allows us to step back and examine the extent to which practices shifted and regrouped into more sustainable long-term configurations – with implications for water-scarcity concerns elsewhere, notably in the United States.

Research shows that people do not experience their everyday use of water as the use of a certain number of litres of a resource, rather their experience is tied up in 'habitual enjoyment of the *services, technologies* and *experiences* that water makes possible' (Allon and Sofoulis 2006, 47, italics in original). That is, people are trying to achieve clean clothes, green gardens, non-smelly bodies, restful surroundings and so on.

Diversity in Cultures of Water

The first point to emphasize is the great diversity in attitudes, behaviours and practices, both within individuals and between different groups of people. This diversity potentially takes us in different directions, both towards and away from water saving. In many (if not most) Australians, watery desires co-exist with a respect for water conservation as an important issue. We love water – for swimming, washing, relaxation – but recognize that we live on a dry continent. Arguably, this apparent contradiction is the basis for our widespread enthusiasms for saving water; we do not have 'electric' or 'petrol' desires in the same way.

A variety of informal water-saving practices were documented during the drought years of the 2000s, encapsulated by the motif of 'the bucket in the shower' (Head and Muir 2007a; 2007b), whereby soapy shower water was collected for use on the garden. These practices (not letting taps run, reducing length of showers, informally capturing grey water in kitchen, bathroom and laundry) occurred both inside and outside the home, influenced by intensive media campaigns and increasingly severe restrictions on outdoor water use. Notable in several studies (Sofoulis 2005; Allon and Sofoulis 2006; Head and Muir 2007a; 2007b) was the intensive labour people were prepared to invest in saving their gardens (see also Chapter 16). This capture of indoor water for use outside helps explain why, contrary to expectations, per capita water consumption around that time in Sydney showed little difference between separate houses with gardens, and apartment or unit dwellers (Troy et al. 2005).

On the other hand, cultures of high water consumption were documented in new upmarket housing estates where people constructed leisure-centred garden spaces to enhance social status (Askew and McGuirk 2004), and among some water tank owners who used a lot of water for leisure (washing boats or wetsuits), hosing hard surfaces and in the garden (Moy 2012).

An important point to note across all these studies is that attitudes do not necessarily map onto practices. Some of the most avid water savers expressed vehemently anti-green attitudes (Sofoulis 2005, 447), drawing instead on a rhetoric and identity of frugality and being anti-waste. Conversely, water tanks could provide a badge of green identity in high-consumption households without necessarily changing practices (Waitt et al. 2012b). A number of studies identified the importance of living under regimes of scarcity (for example in a rural or overseas childhood), in forging lifelong practices of frugality (Allon and Sofoulis 2006; Head and Muir 2007a; 2007b; Moy 2012). As Sofoulis argues, 'practices and values can operate somewhat independently (a feature of adaptiveness as well as hypocrisy!)' (Sofoulis 2005, 451). These findings remind decision-makers not to get too hung up on attitudes, but rather to focus on practices.

Technology Alone Will Not Save Us: The Water Tank Story

The technology of 'Big Water' has been subject to considerable critique in the social sciences (see for example Sofoulis (2005), and various chapters in Troy (2008)), including when it provides problematic friction against which consumers who want to change their practices must push (Lawrence and McManus 2008). But the complex interactions between technology, behaviour and cultural meanings are also evident in the 'smaller' technologies which became more widespread during the drought years – water tanks, shower timers and flow restrictors (Hobson 2006). Candice Moy's work on water tanks in the Illawarra region of New South Wales provides an instructive example (Moy 2012).

After a number of decades of prohibition in urban areas, water tanks were rehabilitated during the 2000s drought. They were heavily promoted and subsidized, and enthusiastically adopted in both New South Wales and Victoria (Sustainability Victoria 2011). Moy's analysis provides the first published post-installation analysis of retrofitted rainwater tanks and their effects on mains water consumption. She compared the mains water consumption of over 7000 households who installed a tank during the drought (for two years before and two years after installation, to smooth out seasonal differences) with that of total household mains water use under a regime of water restrictions. Both populations showed about the same amount of reduction – 10.26 per cent for tank households and 10.8 per cent for the wider community.

This was a puzzling finding as the policy view and the natural expectation is that, even when only fitted with outdoor connections, as most are, domestic tanks are a logical way to reduce the consumption of mains water, 28 per cent of which is assumed by Sydney Water (the water supply authority) to be used outdoors. Interviews and an ethnographic study with a sub-sample of these households identified two distinct sets of practices, summarized by Moy as 'water savers' and 'water users'. The former cohered around practices of frugality, and included a number of people who had grown up in the countryside. Water users were vocal in the importance of autonomy and freedom from government restrictions in their reasons to install a tank, as expressed in the following quotes (Moy 2012, 211):

I can do what I want to do. I'm not governed by government rules.

I think, I can do that, [be]cause it's my water.

It's just that freedom that if I want to hose the concrete, I'm allowed.

Autonomy was also identified as an important issue by Gardiner (2010). Comparing the practices of tank and non-tank households in survey results by Waitt et al. (2012b), Moy also showed that tank households were not statistically more likely than others to undertake water saving practices (turning off the tap while cleaning teeth, only washing clothes with a full load, avoiding the tap running while washing dishes, reducing the length or number of showers, reducing toilet flushes) inside the house. (The first three of the above practices were adopted by a majority of all households in the survey; the latter two were a minority concern.)

The implications of Moy's work are yet to be fully worked through, but it is clear that no technological solution – even a low-tech one – provides a straightforward fix. Rainwater tanks do not achieve water savings in and of themselves, but rather become entangled with social practices and bundles of meaning in assemblages that can both increase and decrease water consumption. The challenge is to get all components of those assemblages ratcheting in the same direction rather than rubbing against each other (Shove 2003). It is worth remembering also that even the 'no-tech' tool of stringent water restrictions – apparently quite effective in driving behavioural change – requires a technological regime of public education and compliance to hold it in place.

Inside Versus Outside

Taken together, this body of research shows that we have made significant, albeit neither universal nor irrevocable, shifts in our outdoor water use. The inside of the home remains a frontier to be conquered for water conservation. There are several things going on here. Outdoor water use is relatively public, amenable to surveillance by both government officials and neighbours, and an obvious first step in terms of restrictions. But also implicit in the lack of restrictions inside the house is the idea that water for cooking, washing and cleaning of humans and their stuff is more essential than water for the non-human life forms of the garden, notably plants. This assumed hierarchy of needs was contested by some garden lovers who thought that everyone should get a 'ration' to use as they saw fit (ration-type campaigns in Melbourne and

Brisbane, which focused on per capita consumption, were among the most effective water conservation strategies of the drought).

Kaika (2005) has argued, using the example of water, that the modern home is constructed discursively and materially as a pure space, distant from nature. Hence the pipes and the infrastructure of supply and of waste disposal are hidden from the householder, at least until something goes wrong (see also Davison 2008). The example of people's interaction with their gardens disrupts this view in several ways. The strong desire to maintain gardens, and the associated labour of water collection, recycling and redistribution, is indicative of a lifeworld consciousness (Allon and Sofoulis 2006) extending well beyond the human and particularly towards favoured plants. Those plants exert agency in the exchange by making visible to humans their desperate need for water. They wilt, dry up and die. We have previously argued that 'it is in the relationship between house and garden that people see, understand and participate in the network of water storage and distribution. Their active engagement with these processes enhances their capacity to manage and reduce consumption' (Head and Muir 2007a, 902). Further, and in contrast to Kaika, they were prepared to

> tolerate 'bad' or 'dirty' nature, within certain limits. The bucket in the shower catches and holds (soapy) bodily wastes rather than insisting they be immediately expunged from the house. Used washing machine water, also containing bodily wastes, goes on to sites of food production. Basins containing dirt washed from vegetables and hands are allowed to sit beside the sink until someone is free to empty them on the garden. (Head and Muir 2007b, 901)

Nevertheless Moy (2012) unearthed considerable resistance to taking dirt in the other direction, that is, bringing tank water inside the home. A number of her interviewees thought tank water was 'dirty', or at least of lesser quality than mains water, and unsuitable for use inside the house (see also Po et al. 2004). Moy argues 'that "dirty water" is only tolerated if its reuse is outdoors. Much greater resistance is met at the prospect of bringing water from outside into the home' (Moy 2012, 214). This partly explained why only 5 per cent of tank households had indoor connections to toilets, washing machines or elsewhere.

Inside the house we encounter norms of cleanliness, for both human bodies and their clothes, that are embedding increasing levels of water consumption in the bathroom and laundry (Shove 2003; Troy et al. 2005; Allon and Sofoulis 2006; Davison 2008). One example is teenagers who may have four changes of clothing a day: for exercise, university, part-time job, and going out at night (Sofoulis 2005). The particular dirt of each context, for example the sweat of sport, has to be removed by washing from both bodies and clothes. And guess who by? The mother in that research case reported doing four loads of washing a day. There is no problem waiting for a full load in households with such high throughput of washing (Chapter 7). More than one shower per day is not uncommon among young adults with active and complex lives (see also Shove 2003).

WHAT TO DO?

Falkenmark (2008) argues for the positive role of water as integrator, as the 'bloodstream' of healthy relationships between human societies and the ecosystems on which they depend. People consciously value the fluid, abundant, life-giving qualities of water. However, if we are going to depend on the dry continent rhetoric to mobilize the populace during droughts, we need to acknowledge the wet continent when it is wet. There are strengths and weaknesses in promoting too close a correlation. People see the connection to water, and respond to it, but they do not so easily notice the environmental costs of the underlying infrastructure, for example the electricity needed to pump it around, which remain high during wet periods. Water supply is not just about water. Yet if adaptive management means responding flexibly to scarcity, it surely also means savouring abundance when it occurs, as long as that does not lock in interactions between behaviour and technology that are difficult to undo later.

Householders who have grown up under regimes of water scarcity, for example overseas, or in rural areas, and older people with a well-entrenched ethic of frugality, have considerable adaptive capacity when it comes to water. This contrasts with the view that the more socially vulnerable have the least resilience and capacity to change. In contrast, generations who have grown up with water abundance and social norms of ever-increasing cleanliness are likely to find it much harder to change.

5. Warmth

Keeping cool and staying warm are basic human needs. Throughout most of human history such needs have been satisfied without the central heating and air conditioning technologies that are becoming commonplace. Heating and cooling of indoor, and even outdoor, spaces is rapidly increasing. By 2009, for example, 87 per cent of households in the United States had an air conditioning unit of some sort, compared with 68 per cent in 1993 (EIA 2011). In concert with this, a narrower range of indoor temperatures is being tolerated than in the past (Healy 2008; Strengers 2008; Biehler and Simon 2011).

The recent history of heating and cooling technology uptake is a story of the intertwined forces of changing standards in building design on one hand, and increasing expectations about instantaneous heating and cooling on the other (Ackermann 2002). Far from being a singularly technological achievement, the management of thermal comfort is a sociotechnical process (Klinenberg 2002; Chappells and Shove 2005; Brown and Walker 2008). The ways in which thermal comfort is achieved during warm weather, for example, have changed remarkably as air conditioning technology co-evolved with changing social norms throughout the twentieth century (Shove 2003). Reliance on artificially cooled air has become more routine. As air-conditioned spaces are increasingly sought, buildings are more likely to be designed and built to be artificially cooled. Rhythms of rest and labour have become less attuned to outside weather conditions and more to artificially maintained indoor climates (Hitchings 2010). Siestas, for example, have declined. Air conditioning draws people indoors during hot weather, off the cool verandas and balconies they may have otherwise occupied, changing rhythms of neighbourhood sociability.

Critics have lamented the 'thermal monotony' becoming embedded in knowledge, institutions, people, buildings and machines:

> The widely varied and, often deeply cultural and symbolic, thermal sensibilities of various cultures have become, and are increasingly becoming, subsumed by an innovative and inventive trajectory facilitated by science – thermal monotony. This is not simply a matter of the achievement of 'optimal thermal comfort' but also, particularly via the effect of standards on the form and content of the built environment, a matter of a reduced diversity in thermally influenced practices and behaviours, much of which are highly cultural in character. (Healy 2008, 321)

Consequences of thermal monotony include 'a sense of disconnection from seasonal change, universalizing notions of corporeal comfort, and unhealthy exposures to substances that circulate within tightly constructed buildings' (Biehler and Simon 2011, 181).

Such cultures of thermal monotony also fuel a sustainability dilemma. Since widespread heating and cooling technologies are powered by fossil fuels, their use is associated with increasing greenhouse gas emissions, contributing to climate change (Chappells and Shove 2005; Hitchings and Lee 2008). Nevertheless, artificial cooling is also advocated by some as a tool for climate change adaptation: a means for humans (and possibly some non-humans) to cope with the increasing incidence of hot weather that climate change itself brings about (McMichael et al. 2006). The resultant dilemma is the focus of this chapter.

Householders are often guided by government policy to consider only immediate household monetary costs and bodily comfort, weighed against distant costs of global greenhouse gas emissions, when choosing heating and cooling technologies in the home. Those seeking artificially cooled air in the Australian state of Victoria, for example, are advised to have their air conditioners serviced, choose an approved energy-efficient model with an adjustable thermostat, and engage air conditioning installation professionals to achieve cost-effective and green thermal comfort. In the neighbouring state of New South Wales, government advises householders to reduce air conditioning use; switch off air conditioners in favour of electric fans, and install ceiling insulation. These recommendations address the thermal comfort sustainability dilemma as a matter of individual consumption. The social licence to switch on home air conditioners is not questioned, except with reference to a moral imperative to reduce greenhouse gas emissions. As with other activities and resources throughout this book – including water, food and clothing – the choice to consume appears sacrosanct.

KEEP COOL

Air conditioning use brings wider social effects, such as blackouts and rising electricity prices. Yet such effects are generally disconnected from household energy policies devised by government as a response to climate change. These unwelcome side-effects present additional hidden dilemmas for householders.

Whether to buy (or turn on) the air conditioner can be understood in the broader context of human vulnerability. In the case of heatwaves, vulnerability is much more than a function of exposure to heat. Rather, exposure to heat is entwined with unequal access to resources, infrastructure, representation and systems of social security, early warning and planning (Ribot 2010). Thus temperature conditions and people's physiological characteristics are transformed into differentiated outcomes during hot weather according to cultural practices, political structures and inequitable distributions of cool air (Klinenberg 2002; Brown and Walker 2008). During the 1995 Chicago heatwave, fragile bodies were not simply malfunctioning in isolation (Klinenberg 1999; 2002). Existing conditions of poverty and isolation were present among those (often elderly) people who suffered heat stress and died alone in their homes. Heatwaves, generating little property damage, can be 'silent invisible killers of silenced and invisible people' (Klinenberg 2002, 17).

During heatwaves, retail sales of air conditioners often dramatically increase. In some episodes of heatwave in Europe and the United States, retail stock has been exhausted, suggesting that many people desire immediate relief from heatwave conditions and view domestic air conditioning as an appropriate means to keep cool (Klinenberg 1999; Salagnac 2007). Yet air conditioning alone does not satisfactorily address vulnerability to heat, due to a range of ways in which residential air conditioners are connected to, and disconnected from, social and infrastructural systems (Farbotko and Waitt 2011). Most importantly, when enough air conditioners are switched on, there is a point at which the existing infrastructure can overload and electricity supply will cease for everyone. This situation, which is common in times of heatwave, means that air-conditioned cool air is inaccessible to everybody affected by the blackout (Kovats and Hajat 2008). Those who have chosen more modest forms of cooling (electric fans and refrigerated drinks) also have their access diminished. Everyone becomes isolated from sources of information about the heat – even on television and radio.

One response to the problem of blackouts during heatwaves is for electricity suppliers to 'gold plate' infrastructure, to cater for greater and greater peak demand. Indeed, many distribution and transmission networks are being upgraded precisely for this reason, encouraged by government energy privatization schemes that reward power companies for 'poles and wires' investments (Beder 2012). An outcome is increasing electricity prices for householders above and beyond the actual costs of transmission and distribution (Strengers 2008). Increasing air conditioner demand in homes fuels the upward spiral of demand for more electricity as temperatures rise and air conditioner installations increase. Capacity at peak periods might therefore be increased, but at ever-increasing costs. The question of reducing demand – asking or requiring householders to restrict air conditioner use – is often anathema to electricity suppliers, who accord householders a moral right to consume and choose as they wish, as long as they can afford it (Strengers 2008). Over time, electricity prices rise from a combination of legislative rewards to power companies for 'gold plating' infrastructure, and decisions made by households with air conditioners to maximize their individual comfort in times of heatwave.

Under this scenario, the rich get cooler and the poor stay hot – and even those without air conditioning must dedicate a larger proportion of their income to electricity bills. Reliance on residential air conditioning also presents other problems. Buildings that have been designed to be artificially cooled with air conditioners can become heat traps when power outages occur, during periods of high demand in hot weather (Klinenberg 2002; Brown and Walker 2008). Furthermore, there seems to be little assurance that infrastructure expansion will eliminate the risk of power blackouts. On the contrary, policy makers are incorporating the risk of power blackouts into their heatwave plans (EuroHeat 2008). Significant here is the widespread assumption that residential air conditioning is a matter of private comfort and individual choice, which cannot be questioned.

STAY WARM

How houses are kept warm varies enormously with geography and wealth. Affluent countries in cooler latitudes typically use central heating, linked to solar panels, gas or water heaters or fossil fuel furnaces, and which are now frequently linked to cooling and ventilation (in a heating, ventilation and air conditioning – or HVAC – system). Such systems, though not without problems (such as when only isolated rooms need heating), are generally considered more environmentally appropriate than electric heating and other inefficient direct heating devices in individual rooms. Other 'green' technologies have pushed central heating horizons further, including geothermal heat pumps, district heating and solar combisystems (Weiss 2003). Nevertheless, similar parameters prevail to air conditioning: efficient technologies are often expensive to install; the rich are able to exercise choice to consume warmth more than the poor; and the vulnerable remain exposed to risks of illness and death in extreme cold snaps – such as in January–February 2012, where in Eastern Europe over 600 people perished, a further 1800 were hospitalized and 75 000 people sought warmth in over 3000 shelters (ABC 2012).

In more temperate climates, how to heat homes is more ambivalent. On the east coast of Australia, where this book's authors live, central heating is rare and homes tend to be built with very open plan spaces, to encourage ventilation in summer. Yet in the depths of winter, overnight indoor temperatures in homes without central heating drop below 16 Celsius – enough to warrant some form of direct room heating. Following steep electricity price hikes in 2012 (themselves attributed to 'gold plating' infrastructure to cope with summer peak demands from residential air conditioning – as discussed above) households made a range of vernacular adaptations. Sales of hot water bottles tripled and hot water bottle covers quadrupled. One retailer said 'old-fashioned methods of keeping snug are on the rise as people try to find the most economical way to keep out the cold and control their energy bills' (quoted in Collier 2012, 3).The St Vincent de Paul Society recorded new cases of the poorest households choosing between eating and heating, or seeking refuge in heated indoor shopping centres rather than leaving heaters on all day at home: 'We are concerned some pensioners and other low-income households are becoming scared to switch on lights and heating because they are so worried about how big their bill is going to be' (quoted in Collier 2012, 3).

Another adaptation was a noticeable return to wood-fired heating, using wood-coal fires in older homes inherited from the Victorian and Edwardian eras. Informal networks of wood fuel exchange sprang to life across urban and rural areas, and within networks of family and friends. Commercial fuel suppliers responded with a wider selection of burning fuels, from the traditional firewood and coal to various 'eco-logs' made from sawdust, wax, industry by-products, and even coffee grounds. In the UK and North America, commentators have debated the sustainability imprint of wood-fired home heaters (Frapart 2010; Gibson 2012), and there is far from consensus on the metrics. Recycled eco-log products do seem to significantly reduce emissions, according to research from the Canadian EPA (Li and Rosenthal n.d.), but the type of wood product burned can mean resulting carbon emissions vary by up to 75 per cent. Exactly how this stacks up against electricity used to heat homes is moot, especially if power generation systems are factored into calculations (for example, coal-fired power plants). Highly variable, and more intermittent, heating practices rule out economies of scale and efficiencies gained by thermostat-controlled central heating. Research has also drawn links between wood-fire smoke and increased death rates, exacerbating asthma and other respiratory diseases (Weinhold 2011).

Further dilemmas stem from the collision of multiple lines of responsibility and ethical conduct: families operate around daily procedures, confront unknowns, and make trade-offs – not all of which are ideal. In our ethnography parents argued with teenage children over heating versus wearing extra clothing:

> Loretta: I suspect that my kids were actually taking little fan heaters into their bedrooms and heating it up and my son... would actually sit in his bedroom with the heater on in winter with a singlet, and I kept saying, Elliot, this is winter you're supposed to be wearing a bit more clothing than just one singlet, that's what you wear in summer. He was wanting to keep it really hot all the time.

Heating using inefficient electric heaters in single rooms may be a necessity when nursing sick children, or for the elderly or frail. Buying wood fuel, hauling it home and lighting an open fire makes immediate the materiality of the carbon being burned – as opposed to the distant, disembodied coal-fired power station enrolled when households habitually switch on (or simply leave on all winter) their electric heater. Open fires run the risk of exacerbating asthma, but conviviality is a clear benefit – families are more likely to gather around open fires than around radiators or reverse-cycle air conditioners. Might the wood fire renaissance be better framed in terms of debates about urban natures and the entangling of non-human others – such as fire – in suburban lounge-rooms? Are urban house fire tragedies more likely? Against risks of burning, children certainly learn to appreciate the power and danger of fire. Our desire to eschew fire danger suppresses mature engagements with it (Pyne 1995).

In other ways switching to wood takes households 'off the grid', lessening reliance on the energy intensification-thermal monotony spiral. But it connects households to other kinds of grids. What are the wider economic, infrastructural and resource use implications if everyone switched to this form of heating? How might shifting patterns of demand alter prices for fuel, availability of resources, or exacerbate downstream environmental impacts in unintended ways? Some eco-logs are made from forest industry by-products; others from recycled sawdust from countries such as Indonesia, where tropical forest loss is a critical ecological issue. Other eco-log products are made from coffee-grounds. Does that mean we need to factor in the chain of impacts related to coffee production as well? Like so many other everyday consumption choices, something as simple as staying warm seems on second thought irrevocably complex.

WHAT TO DO?

More homes could be built to conform to passive home design standards such as the *Passivhaus* standard for energy efficiency in Germany, and the *MINERGIE-P* standard in Switzerland. Such standards require a total building design approach embodying passive principles, greatly minimizing energy requirements to heat and cool buildings. A range of zero-emissions options are also rapidly developing for home design adaptation in hotter climates. Nevertheless, such technologies remain expensive, dependent on expert architectural services, and out of the reach of most people. The vast bulk of populations live within already-constructed buildings, for whom retrofitting is the more realistic option. Other potential actions relate to expectations and norms. Within limits of health and vulnerability, people can come to accept variations in indoor climate as a connection to seasonal variation, rather than a 'problem' to be overcome. Low-tech items include hot water bottles and clothing that improves body-heat retention (or ventilation) without the aid of thermal comfort devices (Head 2012). This remains difficult when department stores and clothing boutiques insist on stocking only generic 'all-weather' clothes, such as business suits – but there are options such as thermal undergarments, and pressure can be applied to employers to allow staff to wear climate-appropriate clothing within the parameters of professional appearance.

Shared, public thermal comfort could be significant in avoiding some of the problems associated with over-reliance on household air conditioning or erratic heater use. Authoritarian measures to reduce residential air conditioning use, by forcing a change in entrenched expectations about individual choice to artificially cool homes, are unlikely to succeed. Air conditioning use in the public imagination, however, may be susceptible to shifting terms of debate. The role of air conditioning in Australian schools, for example, has been debated in the media, viewed by some as a luxury and by others as a necessity. But if private cooling can successfully be mapped to unnecessary cooling, and collective cooling to necessary cooling, householders may be more inclined to seek out public cool spaces during very hot weather, than to attempt to create artificially cooled environments at home. The same situation applies to heating, especially in countries and regions where central heating is not the norm.

Staying cool or warm is one means to reinvigorate sociality, promoting 'animated public spaces and basic resources that pull [people] into the streets' (Klinenberg 2002, 9). Libraries, shopping centres and even hospitals attract people throughout the world during extremely hot and cold weather, because they are spaces where cool or warm air can be accessed for free (Carthey et al. 2009). Nevertheless a much more active attempt to create community-based thermal comfort spaces is needed. Work is particularly required with management of libraries, community centres, shops and other air-conditioned public spaces, to advance, help manage and evaluate practices of collective cooling and heating (Farbotko and Waitt 2011).

Of the homes in our ethnographic study, all located in the warm temperate east coast of Australia, none could be described as sites of thermal monotony. The absence of air conditioning in most of the households was not simply a matter of lack of access to financial resources to purchase a unit or pay for its ongoing costs in energy consumption, although such considerations played a role. Rather, a variety of cultural, economic and material forces shaped the way people achieved thermal comfort: a lack of perceived need; financial costs; negative effects on health; environmental concerns. Bob and Janelle live in a cottage in a seaside suburb and found:

> The house is pretty breezy too so we don't need an air conditioner and we want to spend time like this outside so when it's hot we'll be outside.

Marianna lives in the same suburb and loved the 'beautiful north-easter' that works like a 'natural air conditioner' in her home, so her family felt no need for an electric air conditioning unit. Paul thought that air conditioners were a waste of resources, a source of unhealthy 'false' air, and too expensive. Kylie kept her house so cool with a fan that it felt air-conditioned:

> What we do is because our house is double brick we just keep the windows shut and use a fan and that keeps it cool, and then when it gets dark we open the windows and let the breeze in… People walk in and say, oh, have you got air conditioning?

Knowledge of breezes worked in conjunction with cognizance of the position, materiality and apertures of the home. Householders adjusted their homes and the positions of their bodies in ways that would maximize thermal comfort, as rhythms of wind and temperature varied throughout the day:

> Margaret: I fling every window and door open because I love breezes and I love the house to be really open, so generally speaking we do that, and we put the fans on.
> Elizabeth: I find a breeze, or open the back door and the front door if I'm at home and put that fan on, I get a breeze through the house. Open all the windows, pull the blinds … if there's a sea breeze I don't waste it.

Will and Michelle were also conscious of yearly cycles of hot weather. Very hot weather occurred only a few days annually, weighing against installing air conditioning:

> Will: The old bloke who sold us the house says, no, you don't need air conditioning. I just went, oh. He said, you get a northeast breeze here in summer, maybe 3 days a year you might think you need air conditioning, and he was about right.

Households with air conditioning were also of the view that in the Illawarra region there were few days per year when it was hot enough to benefit from air conditioning. This view shaped the amount of use the air conditioner received. The rest of the time other sources of cooling were used. Marie and her family use electric fans to counter the need for air conditioning:

> During the day we very, very rarely turn it on, only if it's 40 plus because we've also put ceiling fans in all the rooms upstairs. So then that minimizes the need to turn on the air conditioning as well.

Among our group of 16 households in this temperate Australian region (four of which had an air conditioning unit), a cultural expectation of continuous air conditioning in the home was far from entrenched. Where air conditioners were present in households at all, their use was strictly policed in accordance with the prevailing weather conditions. Used only on a handful of days in the summer when it was very hot, they were only switched on for a few hours at a time, and the cool air produced was not allowed to 'go to waste': blinds were snapped shut to keep out warm air, doors were closed, family members were encouraged to be mindful of wasted cool air, and only certain segments of the house and times of the day were deemed worthy of the price (financial or environmental or both) of artificial cooling. Mostly, thermal comfort was achieved by houses being opened and closed according to external temperatures and wind conditions, and by bodies slowing down or shifting location. Air conditioning, when it was present, was not running continuously in the background, but was closely monitored, and always used in conjunction with other ways of keeping cool.

Private spaces therefore ought not to be neglected as important sites of change. Thermal monotony is less visible at the scale of the household. It is common for workplaces and shops to use energy-intensive heating and cooling technologies to achieve thermal monotony, supposedly supporting smooth productivity throughout the day and across the seasons. And while homogenization of home indoor climates is evident in some places, such as Singapore and the United States, there are many places around the world where residential air conditioning and central heating are not normalized, in wealthy countries as well as the global south (Hungerford 2004; Hitchings and Lee 2008). A wider range of practices of thermal comfort might therefore be expected among householders than among employees, given the different types of activities engaged in, and the fact that there remains a significant degree of choice in how best to achieve a desirable indoor climate in many private dwellings. Here is a possible window for policy intervention.

6. Toilets

Householders are being urged to take shorter showers, clean their teeth without running water, and reduce flushing of toilets. But such measures often conflict with treasured ideas about domestic hygiene, and modernity's promise of spotlessly clean bodies and homes (Chapters 7, 15). According to the World Health Organization (WHO 2012) more than 2.5 billion people lack access to any type of improved sanitation facility, and diarrhoeal diseases are one of the main causes of child death worldwide. This chapter explores the dilemmas associated with sanitation, defined as the cultural practices involved in satisfying primal human urges to defecate and urinate, and the safe and sound handling and disposal of human excreta (Avvannavar and Mani 2008). Sustainable sanitation adds a further dimension, responding to ecological pressures of human waste. It involves reduced use of fresh water resources and, ideally, reuse of human excrement as agricultural fertilizer as part of a sanitation cycle (Black and Fawcett 2008; Jewitt 2011). Western households are, generally, participants in unsustainable sanitation systems.

SHIT, AND THE SELF

Our analysis of sanitation dilemmas at the household level starts with the observation that human faecal, urinary and menstrual substances are usually perceived as needing special types of disposal. Unlike dead bodies however (see Chapter 19), whose treatment is frequently influenced by cultural considerations concerning the maintenance of identity beyond death, excrement is almost universally seen as threatening to the self. Faecophilic cultures aside – those rare groups who use human excrement as agricultural fertilizer – the preferred way of dealing with excrement is to remove it from human consciousness and bodily proximity as quickly and cleanly as possible (Hawkins 2006). Human waste products are prototypical objects of disgust. They are viewed by faecophobic cultures as both repellent and abnormal, despite their constant presence in human life (Reinhart 1990; Curtis and Biran 2001; Lines-Kelly 2010). The common reaction to the emotion of disgust is avoidance behaviour; excrement is perceived to be properly dealt with only if it does not leave a visual, olfactory or aural marker, let alone come into contact with human skin (Rosenquist 2005). Human excreta is, by and large, taboo to touch, taste, see, hear or discuss as it leaves the body (Avvannavar and Mani 2008; Bradshaw and Canniford 2010). Such taboos are, according to the World Toilet Organization (2011, 1) the prime barrier preventing development of more effective sanitation technologies globally:

> People are unwilling to discuss the subject for fear of embarrassment ... even at a young age, children are taught that they should not speak such a disgusting subject and quickly we learn that we disempower ourselves when we speak this taboo. But what we don't discuss, we can't improve.

Disgust felt towards human waste products is an evolutionary mechanism for defending the body from pathogens and parasites, with 'the disgust emotion polic[ing] the vulnerable portals of the body' (Curtis and Biran 2001, 29). Yet:

> Levels of disgust vary. Decomposed faeces such as those in septic tanks evoke less disgust than fresh faeces; faeces of babies and family members are more acceptable than those of strangers. People used to defecating in the open find the idea of indoor toilets disgusting, while people with indoor toilets find outdoor defecating disgusting ... These differences indicate that while disgust is a universal primal emotion, it is context-dependent, according to culture, tradition and familiarity. (Lines-Kelly 2010, 14)

In part, disgust towards human excrement is extremely useful. Each society's strict and unwritten rules of how to behave when excreting are associated with cleansing and distancing ourselves from our own waste (Rosenquist 2005). The result is that we are less likely to be exposed to harmful diseases. However, rules of self-purification are linked as much to ethical anxiety as to actual biological danger; perhaps human disgust mechanisms are *too* strong. At the individual level, perceptions of risk associated with excrement tend to be overestimated in comparison to actual risk (Rosenquist 2005). Indeed, 'most of us are far cleaner than we need to be, and most of us have an irrational fear of diseases and contamination' (Hawkins 2006, 58). Thus some disgust mechanisms do not contribute directly to protection from disease. There is no link, for example, between exposure to the sight of another person's faeces and exposure to harmful pathogens. Yet rituals and infrastructures ensuring visually private bodily excretions are commonplace. Bathrooms are often separate rooms, and bathroom doors are often shut and locked while we relieve ourselves (Hawkins 2006). In such systems, the faeces of others are rarely visually encountered beyond those of children or adults in care. Accidental encounters with dirty or unflushed toilets, particularly public ones, are situations from which one makes a hasty retreat.

Rituals of sanitation and privacy are intimately linked to the sustainability of sanitation systems. In some cultures, privacy concerns dictate a social need to relieve oneself not only out of sight, but out of the hearing and smell range of others. Many Japanese women, for example, are highly embarrassed at the thought of being heard by others while urinating. They counter this embarrassment by continuously flushing the toilet to mask urination sounds, using excess water in the process. Addressing both Japanese women's need of aural privacy and their high water consumption, a popular technological device was introduced which creates loud flushing sounds similar to that of a toilet being flushed (the *Otohime* or 'Sound Princess', Avvannavar and Mani 2008). Clean habits and bodily privacy are implicated in much wider cultural concerns than the public health goal of safety from disease (Black and Fawcett 2008). Arguably 'the repugnance attending excreta is at some level dysfunctional, irrational and in need of urgent re-thinking' (Bradshaw and Canniford 2010, 108) if sustainable sanitation is to be achieved.

Even more significant perhaps is the observation that strong taboos about excrement are generally confined to the body and to private space. Less frequently do they extend to public space. In sparsely populated or very hot places, human waste can be readily absorbed by, or quickly broken down in, outdoor areas. In very densely populated areas, however, it is a *lack* of taboo attached to the transfer of human excrement to public spaces that has contributed to extensive disease and pollution, aside from instances where it is safely managed as agricultural fertilizer. Throughout human history, from English cesspools to faeces-filled plastic bags flung out of homes in Calcutta to sewer tunnels emptying into Australian beaches, the risk of exposure to human excrement has problematically been transferred to human and non-human others. This transfer has been conducted in the name of private cleansing, as excrement is shifted out of private and into public domains. There is little difference between piecemeal local methods of excrement removal and efficient centralized sewer systems in terms of our desire to rid the mind, body and home from our own excrement as quickly as possible (Gandy 2006; Hawkins 2006; Bradshaw and Canniford 2010). The public risk associated with accumulated human excrement, however, remains underestimated. Hawkins (2006, 48) has noted that public encounters with excrement can take paradoxical forms: 'horror at the very idea of defecating on the street and resigned acceptance of overflowing stormwater drains and waste treatment facilities pouring raw sewage into the ocean'. Yet no system is without its risks:

No society in the world today deals well with human excreta. At all levels of technical sophistication, damage is done to water, soil, and human health – whether by the pit latrine, the flush toilet, the septic tank/leach field, or, most insidiously and destructively, by the central sewage collection and treatment plant, which creates an unpredictably toxic, and therefore unrecyclable, sludge. (Rockefeller 1998, 17–18)

The 'flush and forget' technology that is ubiquitous in global north sanitation has arguably been the most successful system for eliminating human waste from our private lives. The water closets known as toilets – with a porcelain bowl and water-seal U-bend connected to a sewer – only entered widespread usage after the European sanitary revolution of the nineteenth century. But only wealthy countries have enjoyed this technology as largely ubiquitous, now taking it for granted.

SANITATION: SYSTEMS AND NORMS

It was the flushing toilet that 'elevated water to the position of supreme sanitary agent', contributing to dry systems such as the earth closet, chamber pot, and outhouse falling out of favour (Black and Fawcett 2008, 6). While there are cultural differences in preferred sanitation systems, aspirations toward the water-intensive 'flush and forget' approach are nevertheless common in areas where access to them is limited (Black and Fawcett 2008; Lines-Kelly 2010; Jewitt 2011). Such aspirations are underscored by long-held cultural associations between plentiful water and bodily sanitation. Ancient Hindu scriptures, for example, specify extensive use of water for personal hygiene (Avvannavar and Mani 2008). Toilets have likely contributed to low levels of concern about groundwater pollution from excrement or water treatment quality (Rosenquist 2005; Hawkins 2006; Lines-Kelly 2010). Sewer systems helped to transform 'shit to effluent' (Hawkins 2006, 67) very successfully, placing human waste out of sight and mind:

> Bathroom technologies and techniques minimize encounters with shit. They facilitate the myth of absolute separation from waste, on which mastery of the clean self and its borders now increasingly depends. These technologies provide us with gallons of water, an increasingly scarce public resource, to maintain sensibilities and habits that are fundamental to the organization of our pre-public intimate self. (Hawkins 2006, 59)

Water is the resource that enables the most effective practice of human desires to keep bodily waste out of sight and out of mind as much as possible (Black and Fawcett 2008). It is extremely difficult to shift the norm that water is needed to most properly, effectively and conveniently facilitate the sanitary removal of human waste from the body. Yet such a shift is precisely what is being attempted by water authorities as water conservation becomes a concern. Water-based systems of cleansing ourselves and our surroundings from human excreta are increasingly recognized as unsustainable: they use an unacceptable amount of fresh water, are very expensive, and are characterized by ongoing issues with waste water treatment (Black and Fawcett 2008; Lines-Kelly 2010; Jewitt 2011). Some 20–25 per cent of indoor residential water use in Australia, for example, is due to toilet flushing – about 15 000 litres per person per year (Schlunke et al. 2008). Domestic mains water is typically treated to a drinkable quality throughout urban Australia, regardless of the fact that most of it would 'end up flushed down some kind of drain without going near a tumbler or a cup of tea' (Black and Fawcett 2008, 8). This issue is starting to be addressed in some jurisdictions. Legislation in the state of Queensland, for instance, requires all new houses to be equipped with water-saving devices. This requirement is often implemented by installing rainwater tanks connected to toilets and laundries in newly built houses.

Attempts are also being made to transform Western bathrooms into sites where water efficiency is improved, through a combination of mandated and voluntary measures fostering new technologies and practices around bodily cleansing. Various developments in flush toilet and hand basin technology are designed to both save water and maintain or improve existing standards of cleanliness. Two Swedish municipalities have mandated that all new toilets must be urine diverting, and a 2007 Swedish government proposal aims to return 60 per cent of phosphorus from sewage to agriculture, most of which comes from redesigned toilet systems in apartments (Cordell et al. 2009). Motion-operated or automated taps in public bathrooms appeal to consumers concerned with the cleanliness of surfaces in public washrooms. But they also appeal to the providers of such services, who enjoy reduced use and therefore cost of water, as well as the opportunity to market their services as 'green' (Dodge and Kitchin 2012). A tension between cleanliness and sustainability is also apparent. New toilets with lower flush volumes are being designed to ensure there is sufficient water in toilet bowls to remove all waste without the need for multiple flushes. There also needs to be enough water flowing through drains and sewers to maintain the flow of excrement (Schlunke et al. 2008).

New toilet technologies are often accompanied by expectations of bathroom behaviour change. For example, it is anticipated that toilet users will deploy the 'half-flush' option after urination available in dual-flush toilets, and that male and female toilet-goers will be willing and able to direct their urine and faeces to different parts of the toilet in urine-separating toilets (Schlunke et al. 2008). Yet, there is also an expectation in the sanitation industry that water-efficient technologies will reach a limit when they conflict with treasured toilet practices. In public toilets where large amounts of toilet paper are used to line toilet seats, there is a perceived limit on the reduction in water volume per flush that would be possible. Designers are unwilling to interrupt user desires to keep clean bodies away from direct contact with public toilet seats (Schlunke et al. 2008).

Another example concerns maintenance of comfort rather than health. The higher the ply of toilet paper, the more water is required to flush it down, but there is reluctance in the sanitation industry to question behaviours involving soft or thick toilet paper (Schlunke et al. 2008). Designers of water-efficient toilets have also found that they need to take into account the higher average number of toilets per person in newer homes. Spare toilets are used less; so with lower volumes of flowing water there is more likelihood of blockages by dried paper, causing problems for the plumbing, and the dreaded 'shit gushing up from below' (Hawkins 2006, 57; Schlunke et al. 2008).

Changes in sanitation behaviour, it seems, are unlikely to have a significant effect on water use until there are cultural shifts in taboos surrounding proximity to waste and perceived needs for large quantities of water to flush all signs of it away immediately. In our study of households, many of which subscribed closely to abstract ideas about the need to reduce water use in the home, such taboos were nevertheless strong in everyday practice. Refraining from flushing immediately after defecation was never contemplated, and refraining from flushing after urination was not favoured. Loretta for instance drew on a stereotype of sensitive femininity in expressing disapproval of her family (husband and four teenage sons) not flushing after urination:

> Well, I do [flush after each use] and someone in our house doesn't and I don't know who it is but it's very annoying. And I know that you're supposed to flush every second or third go but I just like them to flush, especially when there are girls in the house because the smell builds up and I don't like it. So there are too many men in here for me to let them not flush.

Josefa was dubious, for health reasons, about water-saving recommendations she had heard such as urinating in the shower. Toilets in her home were flushed each time they were used, except during the night so that others sleeping were not disturbed. For Loretta and Josefa, cultural and health concerns overrode considerations about reductions in water use.

For other households in the study, however, water-efficient technologies were deployed, but not necessarily to 'save water' (cf. Moy 2012; Chapter 4). Rather, they were implemented to justify continued water use in other ways, often to maintain an absence of smell and sight of excrement. Marie, for instance, was satisfied with a potable low-water flush for each use rather than a high-water flush every few uses, because the toilets in her home had a dual-flush mechanism. Evette was satisfied with a grey-water flush for each use because she had instituted a system where used bathwater was stored in the bath and transferred to the toilet cistern by jug as needed by users.

One household in our study had a distinctly different approach to negotiating the tension between cleanliness and reduced water use in the bathroom. Elizabeth and Gerry were avowedly proud homemakers. Over the course of several visits and home tours as part of this study, it was observed that their home was always as neat and clean as a pin. Their commitment to domestic hygiene was high (see Chapter 7). In the context of conversations about toilet use, Elizabeth and Gerry discussed a range of technologies and practices to negotiate their concerns with hygiene on one hand and water saving in the bathroom on the other. With two toilets in their home, they maintained the main bathroom to guest standards, which they conceived as involving a flush for each use. However, Elizabeth and Gerry rarely used the 'guest' bathroom themselves and had different practices in the toilet in their en-suite bathroom, used only by the two of them. In their regular use of this toilet, they followed the mantra 'when it's yellow let it mellow, when it's brown flush it down'. They also coordinated toilet visits, using the toilet consecutively, and therefore flushing only for both. If they discussed their needs to visit the toilet, and tolerated brief visual and olfactory encounters with each other's waste, they could achieve significant water savings as well as relatively quick separation from their bodily waste. They combined such efforts with other water-saving techniques in the bathroom: in-and-out showers; showering less often; water-saving shower roses; a five-litre bucket in the shower to catch cold water running through the shower before hot (then used on the garden); and a one-litre bottle of water in toilet cisterns to reduce the amount of water used per flush. They were proud to report that their water bills indicated a water consumption equivalent to one person or less.

Elizabeth and Gerry had started, in a small way, to re-evaluate both their relationship with excrement, and the resource intensity that currently characterized its disposal. Finding ways to achieve both cleanliness and water reduction in the home, they confronted and altered the disgust felt towards their own and their partner's waste. Human waste was allowed to linger a little longer than usual, albeit only in their private bathroom. Elizabeth and Gerry have started the task of calling private bathroom practices into account, undertaking critical reflection on their personal ethics and practices of dealing with their human waste (Hawkins 2006).

WHAT TO DO?

Infrastructural commitments to water-based systems are hard to change, as are sanitation practices. Improving sanitation for the world's poor remains critical – access to toilets could reduce child diarrhoeal deaths by 30 per cent (WSSCC 2012). Addressing issues of disgust toward human waste, like Elizabeth and

Gerry had done in their small way, is likely to be a central first step to systemic changes to sanitation systems – in both the global south and the affluent north.

The best ecological sanitation (or 'ecosan') systems eliminate water use and water pollution through dehydration and composting, and human waste is transformed into a resource: agricultural fertilizer (Jewitt 2011). Humans consume around three million tonnes of phosphorus each year globally and excrete almost all of this in urine and faeces. Phosphorus is a key component in agricultural fertilizer, underpinning global food production. Yet terrestrial sources are dwindling, and predicted to decline rapidly from as early as 2030 (Cordell et al. 2009). Currently only about 10 per cent of this precious resource is recirculated back to agricultural soils and aquaculture ponds. Most of the phosphorus excreted by urban humans ends up in waterways as a pollutant, or in sewage sludge buried in landfills (Lines-Kelly 2010). The assumption is that effluent-loaded water is only suitable for disposal, not recycling, and that the environment is capable of assimilating such waste, effectively positing waterways and oceans as appropriate dumps for human excrement (Jewitt 2011). Will society reach a tipping-point beyond which such assumptions can no longer be sustained?

Aspects of sustainable sanitation such as handling, containing and reusing excrement clearly conflict with current cultural desires to avoid and dispose of human waste (Rosenquist 2005). But feelings of disgust that such waste often invokes are context-dependent and can be changed, even among fastidious supporters of cleanliness. Human practices of sanitation can be sustainable, as the faecophilic cultures of southeast Asia have long demonstrated. The household is a useful site where disgust with excrement can begin to be confronted, starting with the least disgusting excrement of self and family members. Combined with water-saving practices, this confrontation could begin the long, necessary process of shifting to sustainable sanitation.

7. Laundry

Underlying household cleaning practices, such as laundry, are many complex social and moral assumptions. This chapter discusses changing patterns in laundry technologies and cycles in Western societies. The dilemma of keeping clothes clean is often expressed in simplistic terms of consumer choices between washing machine technologies (Davis 2008; Hustvedt 2011; Katayama and Sugihara 2011) or the chemical composition of different detergents (Laitala et al. 2011). Government policies in some countries include labelling of energy and water efficiency rating of washing machines and regulating the toxicity of detergents. Nevertheless Laitala et al. (2012, 228) argue, 'increased washing frequencies and the amount of clothing we own in Western societies potentially offsets the technological improvements'. This chapter gives particular attention to the taken-for-granted cultural assumptions that sustain distinctions between 'dirty' or 'clean' clothing (see also Chapter 3). Understanding laundry practices and what is categorized as 'dirty' or 'clean' is not constant, but rather constantly changing within and between societies. Keeping clothes clean is exemplary of how everyday norms contribute to household consumption of water and energy, even amongst households that are knowledgeable about and committed to sustainability.

SUSTAINABILITY ISSUES

Worldwide, there are an estimated 590 million washing machines in private households (Pakula and Stamminger 2010). About one third of the world's population uses a washing machine to keep their clothes and other home textiles 'clean'. In the global north, the market for washing machines is almost saturated, while market penetration is steadily rising in the global south. Health authorities now mostly agree that lower washing temperatures can be used in domestic laundering, with the exceptions of vulnerable groups and during epidemics. Recommended washing temperatures have dropped from boiling down to 60 degrees Celsius in the 1980s, to 40 degrees Celsius by the 2000s (Laitala et al. 2012). This drop occurred in parallel with the introduction of new fabrics unable to withstand higher temperatures. But fabrics and washing machine technology are only part of the story of household sustainability.

Table 7.1 Ownership and use of washing machines in private households, selected countries

Market	Ownership rate	Model type	Most frequently used wash temp C°	Yearly number of wash cycles per household (kWh)	Electricity consumption per wash cycle (litres)	Water consumption per wash (m³)	Yearly water consumption wash clothes (kWh)
South Korea	100%	90% vertical	Cold water	208	0.37	140	29.1
Japan	99%	97% vertical	Cold water	520	0.1	120	62.4
Australia	97%	75% vertical	20–40	260	0.34	106	27.6
Western Europe	94%	98% horizontal	40	165	0.95	60	9.9
North America	86%	90% vertical	15–48	289	0.43	144	41.6
Eastern Europe	99%	98% horizontal	40	173	0.97	60	10.4
Turkey	63%	90% horizontal	60	211	1.35	–	–
China	61%	90% vertical	Cold water	100	0.10	99	9.9

Source: Adapted from Pakula and Stamminger 2010, 367

Patterns of energy and water consumption cannot simply be predicted from the design and uptake of new technologies (Shove 2003).

Electricity and water consumption is regionally differentiated because of preferences for top- or front-loading machines. Front-loading, or horizontal axis technologies that wash by compression, are most widespread in Western Europe. Horizontal axis machines are less energy efficient, but consume much less water per cycle than top-loading, vertical axis machines. In contrast, top-loading machines – more widespread in Japan, North America and Australasia – are generally less water efficient than front-loading models, but more energy efficient because they operate well using cold water (see Table 7.1). Often, the lowest minimum programme temperature for front-loading machines is 30 degrees Celsius. As a result, front-loading machines consume more energy per cycle because of the electricity used in heating up water, even on the coolest programme. As illustrated in Table 7.1, electricity consumption per wash cycle is lower in countries where top-loading machines are generally more wide-spread. This global pattern of electricity consumption may change in the near future, because of the rising penetration of front-loading machines into the Asian, Australian and North American markets, the growing number of single households, and the increasing frequency with which people wash clothes (Chapter 3). In the 38 countries studied by Pakula and Stamminger (2010), Turkish households expended the most electricity on keeping clothes clean. They attributed the high value of electricity consumption per wash cycle for Turkey to the relatively large household size of 4.1 persons, and the higher washing temperatures of front-loading machines. The study also showed that Australian, Japanese and North American households run more laundry cycles than those in Europe, with Japanese households the highest. This high frequency was attributed partially to the relatively small load size and short, cold washing programmes. Chinese households ran the least number of wash-ing cycles, pointing to the ongoing importance of manual labour in laundry practices.

TRANSFORMATIONS AND COMPLEXITIES

Social Agendas

Social anthropologist Mary Douglas famously related 'rituals of purity and impurity' (Douglas 1984, 2) to social order and disorder. For Douglas, dirt was understood as 'matter out of place' (Douglas 1984, 40). In making the decision to keep something clean, we 'are positively reordering our environment, making it conform to an idea' (Douglas 1984, 2). Deciding what is dirty is underpinned by taken-for-granted assumptions; cleanliness and dirtiness are never absolute. There are various social agendas that inform the ways in which household cleaning practices are undertaken.

One such agenda is smell. Corbin's (1986) cultural history of smell traced contemporary notions of the benefits of deodorizing or scenting laundry with 'natural fragrances' back to seventeenth and eighteenth century medical discourses. Scents were prescribed to prevent the allegedly contaminating odours from cess pools, cemeteries and marshes penetrating the pores of the body. Today scents and fragrances are still used to cover smells and play heavily into marketing of cleaning products. As shown below in our ethnography, smell has an enormous sustainability impact through everyday laundry practices.

A second agenda is the idea of cleanliness and moral righteousness as the converse of dirtiness and evil. This agenda was not always present. Bryson (2010, 370) notes that early Christians viewed Roman baths as licentious, and 'developed an odd tradition of equating holiness with dirtiness'. Plagues in the Middle Ages challenged this, but led people to the faulty conclusion that 'bathing opened the epidermal pores and encouraged deathly vapours to invade the body' (Bryson 2010, 370). For a further six centuries people did not wash. Only from the 1780s, with the revived fashion for spas and the onset of Victorian values, did notions of hygiene reverse (and dramatically so), equating domestic cleanliness with a naturalized moral social order. And as outlined by Shove (2003), cleanliness was not only next to godliness, but also a means of maintaining social difference along the lines of race, class and gender. People of less social and moral worth were and still are often portrayed as 'dirty'.

A third agenda followed the scientific 'discovery' of the microbe, justifying cleaning as countering the spread of bacteria, germs and disease. One consequence was to reinforce hygiene as central to domestic cleaning practices (Latour 1988). Being a 'good' homemaker came to rely upon caring for household members, through maintaining regimes of hygiene to prevent the spread of diseases – including boiling and bleaching clothes (Sivulka 2001). For centuries wearing separate layers of clothing has also been informed by fear of disease (Vigarello 1998). Underwear prevents contaminants seeping into the body (the rationale behind the medieval practice of wearing the same underwear constantly, without washing it), as well as soaking up problematic bodily fluids.

Who Does the Laundry?

The work of cleaning clothes has conventionally been allocated to women, fashioned as a practice central to sustaining nuclear family life. As Kaufmann (1998) reminds us, the reproduction of gendered identities is bound up in the definition and distribution of the domestic labour of laundering. Keeping clothes clean has become pivotal to the social reproduction of nuclear family life (Cowan 1983). While domestic washing machines have reconfigured the bodily practices of cleaning clothes – eliminating the practices of hand wringing, rinsing, soaping, bleaching and starching – our ethnographic research suggests that in most nuclear family households laundry is typically still considered as both housework and women's responsibility. This was illustrated by John, a 19-year-old high-school student living with his parents and sister in a northern Wollongong suburb:

> Interviewer: Tell me about the laundry routine in your household. Who does the washing?
> John: My mum does the washing most of the time, and my sister too. Sometimes I do it for my clothes only.

Richard, in his mid-20s, lived alone in an apartment and commuted long distance to work in a managerial position; his mother continued to launder his clothes:

> Interviewer: Can you tell me about the laundry routine in your household?
> Richard: Well, because I work fulltime and all that, and I'm really lazy what happens is that my mum gets every Tuesday and Wednesday off work, so generally I would just pile my clothes into one big pile and make sure they're all there, and I will come home on a Tuesday arvo [ed note: Australian slang for afternoon] and magically all my laundry is taken away and then the next day my clothes get returned all nicely washed and folded and everything. It's pretty good my laundry routine.
> Interviewer: So who does the washing?
> Richard: My mum. She doesn't live with me but she comes and does my washing and cleans my unit for me too.

Because expectations of cleanliness have increased, research in the United States and Europe (Coen-Pirani et al. 2008; Kline 2000) has shown that women spend as much time cleaning as their mothers. Furthermore, it is typically women who sort and wash clothes, and purchase laundry powder.

Fears of contamination underpin the process of sorting laundry. Jenny, a 26-year-old student, lives with her partner and two other men in a shared household, yet bears responsibility for sorting the domestic laundry:

> Jenny: My cardigans will only get washed with my stuff. I won't wash them with (my partner) Drew's stuff. Normally it gets split into my stuff and Drew's stuff or it will be like, mine and Drew's t-shirts, then mine and Drew's pants, then undies will go in the t-shirt pile and then my dresses and cardigans will get washed separately from Drew's shit ... I don't split whites and colours. Lighter colours, I never do that. But, like yesterday, I had pretty much all our washing mine and Drew's (clothes), towels and doona [continental quilt] covers. I would put doona covers and sheets all together. I would put all towels together, beach or whatever. Tea towels normally go in with my doonas or my towels. They never get washed with my clothes.

Interviewer: Why wouldn't you wash them with your clothes?
Jenny: Because it's a different type of dirt. I would feel like my clothes would be
  contaminated by the kitchen, and I don't want that.

Methods of sorting vary between households. Sorting may be to maintain categories of colour or fabric, but can also represent social and spatial divisions
such as work and home; partner or flatmate. Sustainability policies around
prosaic activities such as laundry inescapably interact with wider structural
and labour roles (Organo et al. 2012).

Why Wash Clothes?

Laundry practices among our participants were shaped by the practice of
wearing separate layers of clothing, and the distinction between under and
outer garments. In our ethnography, Alice, aged in her 20s and living in a
shared household, assigned underwear to the laundry basket on a daily basis,
underpinned by the enduring importance of personal hygiene and the taken-
for-granted valuing of cleanliness:

Alice: I definitely change my underwear more often than my pants and I'd get more
  wears out of my pants than my underwear.
Interviewer: Why is that?
Alice: I think just basic hygiene practices. I was taught from a young age that,
  that's what you do (laughs). Underwear is a one day thing and then you change
  it and I think with pants unless I have dropped something on them during the
  day or unless I have got really sweaty, I will try and get a few wears out of
  them.

The decision to launder an item is often tied to these kinds of physical descrip-
tions of contact with the skin and/or bodily fluids. Rather than laundering of
garments being informed by germ-related scientific theories, the practice has
subtly shifted and is increasingly 'an exercise in restoring clothes contaminated through contact with the sweaty, smelly bodies of those who wear them'
(Shove 2003, 125). Frequent washing of underwear both rids it of bodily
secretions and restores a feeling of comfort.

Phil, aged in his early 20s and living with his mother, differentiated practices of clothes care between 'school shirts', 'nice shirts' to wear when 'going
out' and 'daggy shirts' – those that he wore around the house:

Undies, I'll only wear undies once, for a day at most, and I won't wear them again,
they'll get washed. If I put on a pair of undies for a day it has to be a clean pair of
undies. Jeans, I rarely wash. I wash them once maybe every two months. Jeans are
pretty okay. T-shirts I'll probably wear two times, it depends on what I do. Like, my
school shirts I wear, on average, twice before I wash them. And they get pretty dirty

throughout the school day, so they have to be reasonably clean. Same with shirts going out, nice shirts going out, I usually only wear those once. Like if I go out, and I know that I'm having a big night, I'll put on a t-shirt and when I get home I'll put it straight in the wash because I don't think that I'll want to wear it after a big night out. But other t-shirts that I wear around the household, I'm not too worried about those. They can have a few wears before I have to wash them... it's usually the shirts that I like least, I wash least.

Clothing care has both a sensual and a social dimension (Pink 2005) (see also Chapter 3).

Deodorizing

Deodorizing clothes by laundering is crucial to maintaining values of style, feel and self-presentation – less a hygienic than a restorative practice. Consequently, ethnography participants spoke about smell as crucial to differentiating between clean and dirty clothes. In response to the question 'How do you decide what to wash?', Victor said: 'It's just about the smell... if it looks really clean but smells dirty, smells of BO [body odour], smells of petrol, well yeah then it needs to be washed'. Similarly, Phil prioritized smell over sight in his categorization of dirt: 'So stink, generally, I'll be looking for first, to see if it smells okay. But then also get it out and have a look at it and see if there's a bit of a beer stain down the front, you know, I'll put it in the wash'. Zara, Theresa and Alice spoke about the 'sniff test':

> Zara: I crotch smell my jeans and if they smell even slightly, they're in the wash. So yeah, I'm a bit fussy.
> Theresa: Sniffing is a pretty sure way of telling something is dirty or clean.
> Alice: I will definitely give it a sniff and give it a once over and have a look at it and see if it's got any marks on it especially under the arms or something.

Unlike in other times and places, where far more ambivalent attitudes have prevailed, in Western cultures, body odours are considered disgusting, particularly in public (Ashenburg 2007). The smell of sweat as 'stink' underpins the classification of clothes that absorb sweat as dirty. The smell, touch, sight and taste of bodily fluids 'come with social and cultural histories attached to them' (Miller 1997, 8). Body odour was a cause of anxiety for ethnography participants, because of how sweat-as-stink threatened to upset codes of cleanliness. Body odour differentiated respectable from disrespectable people. Jenny spoke of her anxiety about clothes smelling sweaty in the presence of others:

> Yep, more about smelling clean, than how it looks on me. I don't care if I have a stain on my shirt it just means I've... you know... been messy that day. It doesn't

bother me at all … If I was already conscious of the fact I smelt that day and someone told me, I'd feel shit. I would be so embarrassed … I guess that's why I take deodorant and perfume everywhere.

Respondents routinely used the emotional terms 'embarrassed' and 'uncomfortable' when they understood the smell of their clothed bodies to violate social norms. Tom recalled how

a guy came into work the other day and it was off-putting, you weren't able to concentrate as well as you normally could have … anything that's out of the ordinary is going to distract you and take your attention away. I mean it is abnormal, but there is kind of a foul context to it as well.

Dangers of contamination modulate into moral condemnation (Low 2006). Moral purity must be achieved through washing clothes (Zong and Liljenquist 2006), and smelling body odour was considered 'insulting and inconsiderate' (according to participant Jenny). Laundry practices are shaped by normative assumptions about dirt, smell and morality. A possible implication is that people will have a large enough wardrobe to wear freshly laundered clothes on a daily basis (Chapter 3).

WHAT TO DO?

Unravelling deeply held social norms and gendered divisions of labour is likely to be a long-term enterprise. Alongside dilemmas over the toxicity of laundry detergents and efficiencies of washing machines are dilemmas arising from laundry as a practice of care and social reproduction. Various social orders are maintained by designating clothes, and people, as dirty, or by consigning cleaning duties to women within households.

Men should do a fairer share of the laundering, and most clothes could be washed less often (Chapter 3). Cleanliness is a necessary part of maintaining health and well-being, but there is nothing natural or fixed about social norms – and many of our current laundering practices may have little to do with disease-prevention. Clothes can be aired rather than automatically put into the dirty basket, and we could better come to grips with body odour and dirt.

For many, though, the shame or embarrassment of failing to live up to what is understood as clean, non-offensive, or fragrant, presents an insurmountable barrier. More gradual cultural change is needed too. Working against reducing consumption of water and energy resources are powerful sensual and emotive norms. People are judged by how they smell in their everyday interactions with others and the use of artificial perfumes has increasingly become the norm. Keeping clothes clean is integral to maintaining the image of a 'good' homemaker, but also a 'responsible' citizen, by not smelling of body odour. Such notions increasingly collide with the imperative to become a 'responsible' citizen by reducing energy and water consumption.

8. Furniture

Of all the things we cram in our homes, among the bulkiest and most expensive are items of furniture. Using the home contents of the co-authors of this book as a guide, the average middle-class home today contains three or four beds, two bedside tables, a wardrobe and chests of drawers for each person, one or two couches and armchairs, a coffee table, two desks and desk chairs, between three and eight bookshelves, a dining table and six to ten dining chairs, an outdoor table with four to six outdoor chairs, a TV cabinet, hi-fi and/or LP cabinet, CD/DVD storage units, and a host of 'miscellaneous' extras (entrance hall tables, stacking tables, trolleys, side tables, pianos). This chapter considers the sustainability dilemmas of such furniture, and considers one example in more depth – mattresses – where a range of complexities transpire.

SUSTAINABILITY ISSUES

For all the space it takes up in our homes, furniture is, curiously, not so much at the forefront of sustainability debates or policy making. A major UN report on sustainable household consumption for instance contained detailed information and best-practice examples on a range of goods and activities (heating, hot water, food, energy, transportation, labelling) but said nothing about furniture (UN DESA 2007). Similarly Wright et al.'s (2009) otherwise comprehensive book on the sustainability impacts of the stuff in our home had no index entries for furniture, beds, mattresses, tables or chairs.

One reason for this absence may be that where data is available, it suggests that the contribution of furniture to sustainability problems is slight: according to the Australian Conservation Foundation (ACF 2007) furniture accounted for less than 5 per cent of average household greenhouse gas pollution, barely 1 per cent of water use, and a similarly tiny fraction of overall ecological footprint.

Yet this may be misleading. Tukker et al. (2006) noted that furniture is unwittingly overlooked in most studies that approach household sustainability by accounting for energy, carbon or water use. Unlike household appliances, which can be directly measured for energy or water use, data on furniture is inconclusive or non-existent, which is not to say that there are no resource use implications, nor downstream and upstream lifecycle impacts. Tukker et al. (2006) estimated that furniture accounted for 27 per cent of household waste – the highest category.

Other impacts are related to the range of materials used. According to the European Commission (EC 2008) the different materials used in contemporary furniture production, by value are: 26 per cent wood-related materials (hard and softwoods, veneer, MDF, particleboard, plywood); 17 per cent components, 12 per cent metal, 10 per cent hardware, 9 per cent plastics, 7 per cent textiles, along with smaller percentages of leather, glass, marble/stone and rubber (though this varies geographically; Swedish furniture industry figures, measured by weight, put the figure for wood-based materials at more like 70 per cent). An immediate challenge presented is that for even a single piece of furniture it may 'be difficult for manufacturers to comply with environmental requirements for all materials used' (EC 2008, 6).

Unlike other household products, where a disproportionate contribution to the environmental imprint stems from either their assembly/manufacture or post-purchase use (for example, televisions, Chapter 13, and clothing, Chapter 3), for new furniture the impacts 'stem mostly from the production and treatment of the raw materials used in the manufacture, rather than from the production of furniture itself' (EC 2008, 7). Wood used in the manufacture of furniture has been linked to international trade in virgin forest products and unregulated clearing practices (Tukker et al. 2006). Uncontrolled logging for the furniture trade is a major contributor to loss of habitat and biodiversity, erosion and soil degradation (EC 2008). The use of tropical woods from endangered rainforests has been a particular source of criticism (Handfield et al. 1997), reflecting concern over these globally important biomes, as well as past experiences with the over-exploitation of teak (from Asia) and mahogany (more precisely *Swietenia mahogani*, from Cuba, Haiti and the Dominican Republic), the latter of which was 'irremediably extinct' within 50 years of its discovery (Bryson 2010, 174). Meanwhile, the production of fibreboard, plywood and particleboard requires heat, pressure and adhesives (such as formaldehyde resins, melamine, epoxy, polyurethane and ethylene vinyl acetate), producing dust, hazardous waste and emissions and risking the health of workers (Schlünssen et al. 2001; Vaajasaari et al. 2004).

Nevertheless, wood can be sourced from forestry activities operating under stricter controls. 'Best practice' in the industry recognizes wood as a renewable resource, and promotes higher percentage use of wood in manufacturing, compared with metals and plastics. Metals, in contrast to wood, are not renewable, and are linked to contamination of local water sources, noise and dust emissions during mining, higher levels of energy consumption in production, release of heavy metals through wastewater emission of fluorine, nitrogen oxide and sulphur oxide, and through emissions of other compounds when undergoing surface treatment (galvanization, painting, lacquer, enamelling) (EC 2008). Such burdens can be reduced through increased use of recycled metals in furniture production (especially aluminium). Accordingly, the European Commission now recommends that furniture design should include metal parts that can be easily removed for selective collection, to facilitate recycling. Metals are also durable and are often favoured for precision in design.

Plastics used in furniture production are made from natural gas or oil, and are hence linked to the environmental, economic and social problems of those industries. As with metals, use of recycled plastics has been encouraged. Textiles, meanwhile, are linked to a whole host of other raw materials processing and production problems, and environmental and ethical concerns, depending on whether cotton, polyesters or leathers are used (see Chapter 3 on clothing).

A persistent dilemma is that wood, although preferable, is distinctly expensive in comparison with metal and plastic. An ever-increasing share of the market for furniture is controlled by cheaply made furniture from low-cost raw materials and cheap labour countries (notably China, Vietnam, Indonesia and Mexico) (Buehlmann and Schuler 2009), with often higher plastics and metals content. The environmental ethics of furniture consumption is therefore cut across by social justice concerns such as the availability of affordable furniture for the masses, and the environmental and social conditions of production.

The materiality of furniture is also significant. Wooden furniture has one of the longest lifespans of all household items, and reuse habits are well established. Furniture has traditionally been expensive, valuable and long-lasting, spawning intimate links to the antique and second-hand trades. The long lifespan of furniture has led to it being seen as an important sustainability parameter: 'a product that can be used for a longer period of time will need to be replaced less often, which has an overall positive effect on the environment: less usage of raw materials, less pollution related to production, and less waste' (EC 2008, 11). Furniture – especially that made of wood – breathes and warps, marks and matures, inviting owners to cherish it more.

Nevertheless, this is premised on furniture being well-built and made to last in the first place. Increasingly, furniture is incorporated into 'fashion cycle' logics of cultural industries, rather than being seen as longer-term infrastructure for households (Leslie and Reimer 2003). Whereas furniture purchase was once dictated mainly by price and quality, nowadays it is 'affected by elements of feel good factor, pride of ownership ... furniture is no longer perceived as a functional item, rather it is merchandise ... form[ing] an integral part of the lifestyle and home' (Ratnasingam and Ioras 2003, 233). Furniture is replaced and disposed as waste more frequently, either because of poor quality, or because fashions come and go.

Such trends have accelerated with the advent of 'fast-furniture' superstores, known as 'category killers' for their annihilation of small, local, independent manufacturers and retailers (Sampson 2008). Megastores such as Costco and Walmart commission cheaper products using a higher proportion of metals, plastics and low-quality particleboard and plywood materials. Such retailers are dedicated to lowest price-point competition for furniture; they dedicate a higher proportion of budget to advertising, technological supply and distribution systems.

Others, such as IKEA, have traded sturdiness of construction for enhanced design and aesthetics, positioning their brand in relation to regional furniture-making traditions (such as the 'simple Scandinavian furniture aesthetic') (Hakansson and Waluszewski 2002). Furniture is also increasingly linked to homewares, fittings and other shorter-term purchases, all of which are governed by fashion-cycle and disposability logics. Bathroom and kitchen fixtures are especially prone to this (see Chapter 12).

Such trends are not the exclusive domain of cheap manufacturers: in some exceptional cases furniture design has itself fused with elite cultural industries, emphasizing design-led innovation, marketing and styling of products, customization, consumer 'self-actualization' and hand-crafting (Elliot et al. 1996; Leslie and Reimer 2006). The dilemma – as the success of Danish furniture design has shown – is that well-designed, highly-aestheticized furniture can be finely crafted, becoming highly cherished well beyond its own generation, but remains expensive. Other factors influencing furniture lifespan include 'fitness for use' (perceptions of usefulness, function, comfort and its contribution to a healthy environment for the user), ergonomics (itself governed by national standards, which can render old office chairs for instance obsolete, even if well-made), repairability and availability of spare parts (EC 2008).

Above all this, sustainability issues for furniture depend on household practices. Because of its usually longer lifespan, furniture is one of the first household product industries to be affected during economic downturns (Ratnasingam and Ioras 2003). Indeed, since 2008 furniture sales have declined, 'affected by volatile demand across the housing industry, along with fluctuations in disposable income, consumer sentiment and consumer tastes and preferences' (IBISWorld 2012). The positive spin on this is that during tough times households display a greater propensity to hold onto existing furniture, to swap, gift or reuse furniture, or purchase second-hand furniture. In our ethnography those households most likely to purchase second-hand furniture were those on the lowest incomes, or those, such as Paul, who made conscious choices to 'live simply':

As you can see around here the furniture isn't exactly Adams or Sheraton or something like that. Everything is from the op shop [charity shop] or whatever, including all the clothing. I don't think there's much where I allow great luxury in life at all and that's a set purpose, because I simply want to live simply.

Even during boom times, much furniture is retained by households for future possible reuse – its implicit use value encouraging storage rather than ridding, even given furniture's bulk. As Gregson et al. (2007) showed, households save, stockpile and accommodate – even at the risk of cluttered basements, spare rooms and attics.

More deeply too, furniture saving, gifting and reuse is central to the formation of individual and household identities, such as when children move out of home, inheriting their parents' old furniture, pots and cutlery (Gregson 2007). Furniture gifting and reuse marks a stage in people's lives: leaving home and moving into young adulthood, the passing of parenting, moving to a phase of living without kids. Hence Julie, one of our ethnography participants, marked her daughter's moving out of home by giving away the bulk of her own furniture – thus creating a time to buy a new round of furniture with which to remake her post-parenting home space. Another point of tension in our ethnography was what to do with old furniture passed down from recently deceased parents – the emotional meanings of material things for loved relatives rendering furniture as impossible to discard. Furniture creates feelings of homeliness (Dowling and Power 2011), especially when given by or handed down from loved ones.

These are ostensibly cultural phenomena rather than practices driven by sustainability concerns, but they are related to sustainability burdens and dilemmas: how much stuff should be stored to keep for when the kids eventually move out? How many furniture items are purchased with future lifespan – including designation as 'heirlooms' – in mind?

In our ethnographic work, quality was a consistent factor driving furniture purchasing among environmentally committed households – suggesting that families have not universally bought into the replacement/fashion-cycle logic. As George articulated:

Yeah, we don't throw stuff out because the fashion changes. There's got to be a good reason for it. And we don't throw stuff out because we might get to a house one day and there might be that extra room. We've still got some of the stuff we got given from ages ago. Annoying as it is, we don't just get a whole new bunch of furniture because it's not, I guess... we hold onto stuff for a long time.

Julie similarly pointed to the quality of her couch and armchairs: 'See they're built to last, they're built to last. I won't be chucking them out. I've got a sister who changes the lounge suite because she's tired of the colour. I don't get tired of the colour'. Ethnography participants gave away furniture items to charity stores or left them 'on the front verge' for others in the neighbourhood, rather than re-sell items at garage sales or on eBay. They also spoke fondly of buying (and cherishing) certain types of 'special' furniture items – chairs, bedheads, tables – from charity shops.

Such practices arguably shed new light on other kinds of sustainability dilemmas. Despite larger homes increasing the energy (and to some extent, water) imprint of households, there were gains to be made in terms of the capacity for greater accommodation and storage of long-lifespan items such as furniture – facilitating possible reuse (Dowling and Power 2011). In our ethnographic work with extended family homes furniture was sometimes doubled-up within large structures (especially in kitchens), but was also stored en masse, gifted, shared and reused across multiple generations. The same went for wardrobes – that item of storage furniture par excellence – and for garages, converted from car shelters to intricately organized storage spaces for furniture, appliances and homewares. Wardrobes nowadays are substantially larger than in previous generations, encouraging stockpiling and excessive consumption, but also facilitating higher rates of retention for future reuse over short-term disposal. Nevertheless, how many stored furniture items (and things stored in large wardrobes) actually get reused? Are they eventually discarded too – their stockpiling within large homes merely delaying the inevitable? This was the case for one participant in our ethnography, Julie:

> The garage is full of furniture that doesn't fit in here and I'd had a dummy spit and said, listen, why are we keeping this stuff, it doesn't fit. So yeah, just sell it. If it sells, lovely, if it doesn't well, we'll take it to the recycle depot and they can sell it to folks that have less income than us.

All these issues come together in one vital – if often forgotten – piece of household furniture: the mattress.

THE MATTRESS: SOME THOUGHTS TO WAKE UP TO

The mattress is the article of furniture that you are likely to spend most contact hours with, given sleeping occupies almost a third of our time. For much of human history, beds have been (and continue to be) among the most valuable items of furniture in the home. Beds, like other substantial pieces of household furniture, are infrastructural investments, not disposable items.

Despite their relative invisibility, mattresses have been a constant source of unrest. Only recently in human history have they become stable things, largely forgotten under the sheets. Prior to the introduction and widespread dispersal of inner-spring mattresses (in the 1930s) and then waterbeds (in the 1960s), most mattresses were made and remade from straw, and other proximate or affordable materials including sawdust, sea-moss, wood-shavings, cotton, hair, wool, feathers and down.

Almost universally, the Western norm is now towards buying mattresses new rather than second-hand. Where babies are concerned, this extends further, with official recommendations to new parents to purchase new mattresses, irrespective of the utility of second-hand ones, to reduce risk of Sudden Infant Death Syndrome (Chapter 1). Mattresses are a prime site where perceptions of cleanliness and dirt are manifest in relation to material products (Chapter 7). So much so that nanotechnology is now being introduced into mattress design to introduce a 'self-cleaning' effect – preventing dirt and liquids from sticking to the mattress surface (Cordella et al. 2012). Such concerns are not new, but have their origins in centuries' old associations between beds, dirt and disease.

Over 67 million mattresses are sold annually in the European Union (Cordella et al. 2012) – an industry worth €4.9 billion. Industry research in the EU suggests that half of all replaced mattresses are disposed of via landfill annually; the balance is sold on the second-hand market, given away or stored (Glew et al. 2012). As much as 10 per cent of all landfill is comprised of discarded mattresses (Cordella et al. 2012). Unlike most other types of furniture, which are intended to be long-lasting, the recommended lifespan for mattresses is eight to ten years. Tightening fire, health and safety regulation increasingly renders mattresses obsolete, regardless of the amount of wear and tear. The average double mattress has a carbon footprint of 80 kg of CO_2-e, 44 kg of which is derived from the foam and fillings, a further 29 kg from metals and 14 kg from the textiles used in production (FIRA 2011).

Two key dilemmas for households are whether new mattresses chosen are manufactured with biomaterials – fibre crops, animal hair, or wood – generally more sustainable than those manufactured with petrochemicals; and methods of disposal for old mattresses.

Glew et al. (2012) performed a comparative lifecycle assessment of a high-quality 'natural fibre' and a polyurethane foam pocket spring mattress, over a lifetime of ten years. Recycling and reuse of mattresses, regardless of construction materials, can reduce greenhouse gas emissions by 90 per cent in comparison to landfill. Mattresses manufactured from biomaterials emit less greenhouse gases over a lifecycle than those manufactured from petrochemicals. Glew et al. (2012) concluded that the use of natural fibres and recycling of mattresses offered opportunities to reduce thousands of tonnes of CO_2-e. Given such findings, why do most manufacturers continue to opt to produce primarily polyurethane mattresses? Glew et al. (2012) estimated that only

5 per cent of mattresses contain natural fibres. The answer is bound up in consumer preferences, and the politics of mattress recycling.

Industry consumer market segmentation research commissioned by the International Sleep Products Association sought to better understand why people do, or do not, buy a mattress (*BedTimes* 2009). Results of their survey of 1800 American adults (conducted in September 2008) suggested five distinct market segments: 'mattress involveds', 'sleep sufferers', 'healthy and content', 'brand selectors' and 'apathetics'. 'Mattress involveds' was the market segment most likely to identify themselves as either 'green' or 'ethical' consumers. This group was characterized as professional, younger people with high incomes. 'Market involveds' conducted extensive research before making a purchase, saw a mattress as integral to their health, and were willing to pay a premium price. Their purchase was often motivated by a doctor's recommendation. Ecolabelling was most likely to inform this segment's purchase decision. While they generally followed the shortest replacement cycle of 6.9 years, they were also most likely to reduce their greenhouse gas emission, expressing a willingness to pay for recycling their old mattress.

In contrast, 'apathetics' had the longest replacement cycle, of around 13.5 years. This segment was most frequently comprised of men on lower incomes. They typically decided to buy a mattress as a reluctant necessity, such as when they were divorced. Price was the most important factor, rather than brand, warranty, ecolabelling, biomaterials, or perceived health benefits from improved sleep. While 'apathetics' reduced their contribution to greenhouse gas emissions by retaining their mattresses for the longest period of time, they were generally least likely to be willing to pay a small fee to recycle their old mattress. For most, sustainability was not a primary factor in the decision to purchase a new mattress. Furthermore, the cultural norm amongst all but one market segment was to dispose of old mattresses in landfill rather than pay for mattresses to be recycled.

Landfill remains the most widespread disposal practice for mattresses. Nevertheless, some 85 per cent of the materials pulled from mattresses are salvageable for recycling, including reclaimed steel, cotton fibre, polyurethane foam and wood. However, the low market price for recycled components has worked against most private investors starting mattress recycling businesses. The recession triggered by the 2008 global financial crisis worked against private investment because the sluggish economy further lowered the market price for recycled materials. Our ethnographic research also suggests that knowledge about how and where mattresses can be recycled is poor.

Of the few mattress recycling businesses that are in operation, the majority are run by not-for-profit organizations. Most use manual methods to pull apart mattresses, and are estimated to recycle anywhere from between 4000–20 000 mattresses per annum. The financial viability of mattress recycling operations relies upon charging a per-piece disposal fee that may vary between US$6–15 per mattress. Social enterprises such as the Project for Pride in Living Industries, begun in 2008 in Minneapolis, assist low-income families and provide job training for socially marginalized groups including the homeless and migrants. Through the combined capacity of facilities in Oregon and California, the St Vincent de Paul Society has in this way become the largest North American mattress recycler.

Things are changing, belatedly, through stronger state regulation – though not without contestation. An increasing number of American states have introduced bills to curb the flow of mattresses to landfill, making manufacturers responsible for collecting and recycling of their used product (a cause of some alarm for the International Sleep Products Association). The dilemma nevertheless persists that mattresses – a key source of anxiety about dirt and odour, even with new materials and technologies – are one of the most rapidly replaced items of everyday household furniture.

WHAT TO DO?

Furniture can be bought, swapped or shared second-hand wherever possible, or if not, repaired, reupholstered, or bought new made of wood, if its source and forestry management practices can be assured. Classic items of furniture are often those with timeless style, or that live past fashion-cycles to become retro, or antique. Furniture is an infrastructural item in the house – it was once very much perceived this way, which in turn encouraged higher quality production and craftsmanship, and fuelled greater levels of reuse, handing down and circulation outside the cash economy. New furniture can be examined for sturdiness of construction (improving the chance that it can be on-sold, given away or reused), and for ease of dismantlement (encouraging recycling of wood and metal parts).

Meanwhile, what about mattresses? It is very difficult for those with sufficient means to imagine behaving outside the norm of buying new rather than second-hand mattresses. Unless economic conditions deteriorate markedly, that is unlikely to shift soon. Mattresses are normally hidden from view in bedrooms, and we are unlikely to be condemned by a glance or statement that the mattress we have purchased is 'bad' for the planet. The bigger question is how to resolve the deep contradictions between an aware market segment who exercise choice and care over purchasing and recycling decisions (yet who replace mattresses more frequently) and an apathetic segment that make their mattresses last longer, but who do not recycle, and are unwilling to pay more for better-made, lower-footprint items.

9. Plastic bags

Few of us get through a day without encountering a plastic bag – taking or refusing one at a shop, reusing one that we had taken previously, carrying things, passing them on, storing them. Plastic bags are useful for wrapping food and other wet things, lining rubbish bins, and wrapping delicate materials. They are popular because they are 'functional, lightweight, strong, cheap, and hygienic' (DSEWPC 2012). In parts of Africa they are even bound together into makeshift footballs, in informal settlements where commercially produced footballs are prohibitively expensive.

That popularity has also made plastic bags ubiquitous, in truly staggering numbers. Australians used 3.9 billion bags in 2007 (a reduction from 6 billion in 2002) (DSEWPC 2012). UK estimates are 10 billion bags in 2006 (Edwards and Fry 2011). Before recent bans on plastic bags in China, it was estimated that 3 billion were produced there each day (WorldWatch 2012). It is impossible to know how many are produced worldwide; estimates vary between 500 billion and 1 trillion (1 million million) bags per year.

Although some of the arguments here relate to plastic bags more broadly, the particular focus of this chapter is a subset of that category, the single-use plastic bags with integrated handles provided by many supermarkets. Most of these are made of high-density polyethylene (HDPE). They are the ones commonly seen wrapped around sea mammals and birds, in visuals associated with campaigns against the plastic bag (e.g. www.banthebag.com.au).

Scaled up, the prevalence of plastic waste does indeed contribute to significant problems in waste disposal, threats to wildlife and fossil fuel use. But it sometimes seems that this mundane bit of material culture gets more than its fair share of environmental attention precisely because it is constantly in our faces (and our kitchen cupboards). As Gay Hawkins (2006, 21) describes:

> All those years of public campaigns instructing me to resist the easy convenience of plastic bags and do my bit for nature. This training does not mean that I have completely eliminated plastic bags from my life, but it has meant that my relations with them have become more complicated.

And there is no doubt plastic has a bad press. Participants in our ethnographic studies were very aware of plastic bags and bin liners. One couple said they felt 'embarrassed' about their use of bin liners in different rooms. When asked why they felt bad, the reply was:

> Jan: All the plastics. Don't like all the pollution.
> Josefa: We try not to buy much plastic stuff at all.
> Jan: Just being aware of where it all ends up basically, down in the ocean. So I end up picking up rubbish off the beach all the time when I go swimming and there's lots of plastic bags.
> Josefa: When we go for walks we pick up all the plastic rubbish. In fact a lot of people around here do. Like, all the locals really take very good care of the beach, don't they?

In this example household plastic bag use was connected to littering on the beach. Many coastal residents were very conscious not only of direct littering on beaches, but of the fact that when it rained, a lot of rubbish was flushed down the drainage systems to the beach. Even though plastic bags still featured in their lives, people felt regret for their actions, and made tangible connections between practices inside the home and wider ecological flows and consequences. Conversely, communities who have banned plastic bags are very proud of this fact. Coles Bay, at the entrance to Tasmania's spectacular Freycinet Peninsula, claims to have been the first township in the world to ban plastic bags in shops (in April 2003). It beat Huskisson, on the New South Wales south coast, by six months – who nevertheless proudly claim to be 'Australia's first mainland plastic bag free town', announced by a 'welcome to' sign at the entrance to the town. Such proclamations are in part overt revisions of a community's relationship with plastic – but are also another way to express pride in their 'pristine' coastal location, the coast being the ecosystem perceived to be most threatened by plastic bag waste.

TRYING TO USE LESS PLASTIC

A variety of measures have been adopted to reduce production and usage. Bans exist in South Africa, Kenya, Rwanda, Tanzania, Uganda, Italy and France. Levies in Ireland and the Australian state of Victoria have been shown to significantly reduce usage (Lewis et al. 2010). The relative merits of voluntary actions by retailers and government regulation continue to be hotly debated, although discrepancies between consumer attitude and behaviour suggest that sustained change will require some form of regulation. For example Lewis et al. (2010) report that despite many people responding in surveys that they often or sometimes try to avoid plastic bags when shopping, 67 per cent of supermarket transactions still involve a single-use plastic bag. In our survey, 62 per cent of respondents claimed to 'always' or 'usually' take their own bags when shopping (Waitt et al. 2012b). Most of us have had the experience of leaving our green bags at home or in the car, or just popping into a shop for one or two items and ending up in possession of a plastic bag.

WHICH IS THE BEST PLASTIC?

How do single-use plastic bags compare with the other options? Comprehensive work on this has been undertaken by Helen Lewis, Karli Verghese and colleagues at Melbourne's RMIT University, and the remainder of this section draws heavily on their published findings (Lewis et al. 2010). They argue that while lifecycle assessment provides a crucial baseline for decision-making, the broader social context of bag use – approached by qualitative research methods – is just as important.

They compared the bags that would be needed for a household to carry 70 grocery items home each week for 52 weeks. The streamlined lifecycle they modelled included 'the environmental impacts associated with raw material sourcing and production, manufacture of the bags and their disposal at end of life (i.e. landfill, recycling, compost or litter)' (Lewis et al. 2010, 149). The seven carry bags that were evaluated were:

- a single-use HDPE bag;
- a single-use HDPE bag with recycled material;
- a single-use biodegradable plastic bag (commonly made of corn starch);
- a single-use oxo-degradable plastic bag (commonly HDPE combined with a small amount of prodegradant additive);
- a single-use paper bag;
- a reusable 100 per cent recycled PET (polyethylene terephthalate) bag; and
- a reusable PP (woven polypropylene) bag.

The latter two are commonly referred to as 'green bags', although they are not always green (indeed, they are increasingly coloured and 'branded' as part of marketing campaigns, seen as a new opportunity to more permanently mark brand names and logos onto bags used by consumers – effectively advertising repeatedly for free).

The environmental impact categories used in the analysis were global warming, photochemical oxidation (smog), eutrophication (nutrients released to waterways), land use, water use, solid waste, fossil fuels and minerals. A common finding of lifecycle assessments is that different items are differentially better or worse depending on which impact is of most concern. We do

Table 9.1   Environmental impact indicators for carry bags

Impact category	Unit	HDPE plastic bag 100% virgin	Biodegradable bag	Paper bag	Reusable PET bag
Global warming	kg CO_2-e	7.52	9.19	44.74	6.47
Fossil fuels	MJ surplus	19.93	9.96	44.77	6.46

Note: Analysis standardized to a common 'functional unit', defined as 'the number of shopping bags consumed by a household to carry 70 grocery items home from the supermarket each week for 52 weeks'.

Source: Simplified after Lewis et al. 2010, Table 1.

not explore the full details here, but rather highlight three issues that connect most closely to household dilemmas.

First, and consistent with previous studies, the researchers found that reusable green bags had lower environmental impacts than all the single-use bags (Table 9.1). But this outcome is highly sensitive to the number of times the bags are used. Because of the greater amount of embedded energy and materials in their production, green bags have to be used more than 50 times for this benefit to be realized. Indeed, 'if a reusable PP "green bag" is only used 52 times (weekly for a year) instead of the assumed 104 times (weekly for two years) then its impact on global warming is higher than the impact of each of the single-use bags, except the paper bag' (Lewis et al. 2010, 149). Green bags do not provide benefits by just sitting in the boot of a car, or in the kitchen cupboard.

Second, and again consistent with other studies, single-use paper bags are worse in most impact categories than single-use plastic bags, and worse in all the categories than HDPE bags. For most categories this finding remained unchanged if the bag was assumed to be used twice. Forestry operations necessary for paper bag production have high impacts on eutrophication and land degradation. 'Paper bags generate more greenhouse gas emissions than single-use plastic bags and reusable bags because the manufacturing process is more energy intensive and because of the larger mass of material used over a 2-year period' (Lewis et al. 2010, Table 3). Water and chemicals used in the manufacturing process contribute to high levels of waterborne waste.

Lewis et al. (2010) note that packaging materials such as paper can appear more 'natural' or 'renewable' than plastic, even when the lifecycle analysis evidence demonstrates otherwise. This illustrates again our contention, discussed in other parts of this book, that overtly 'green' behaviours do not necessarily stand up to detailed analysis (see for example, eco-clothing, Chapter 3). Paper bags have less impact in the litter stream because they decompose more quickly than plastic, and pose less of a threat to wildlife. This finding leads to the third implication that place matters in the dilemmas of sustainability. Communities may make different decisions depending on which environmental impacts are most important in their locality. For example, Lewis et al. (2010, 158) suggest that 'a retailer in a coastal area could decide to place a higher priority on litter impacts and therefore to offer paper bags (for sale or for free) as an additional option'. As the Coles Bay and Huskisson examples emphasize, communities that depend for their livelihoods on coastal tourism have strong incentives to work hard at getting rid of plastic bags, to maintain both the image and the reality of clean beaches. In the general stores in both towns, calico bags are offered for sale as an alternative.

Comparable findings to Lewis et al. (2010) were made in a recent UK study. Focusing on global warming potential, Edwards and Fry (2011) also found that HDPE bags have the lowest impact, particularly when reused as bin liners. (They estimated that 40.3 per cent of HDPE bags in the UK are reused in this way.) 'The paper, LDPE [Low Density Polyethylene], non-woven PP and cotton bags should be reused at least four, five, 14 and 173 times respectively to ensure that they have lower global warming potential than conventional HDPE carrier bags' used in this way (Edwards and Fry 2011, 61). This is a notional reuse rate only; most paper bags, for example, are not strong enough to be reused four times. The biggest difference between the two studies is the much higher rate of green bag reuse needed to have equivalent impacts to HDPE (>50 in Lewis et al. 2010 and 14 in Edwards and Fry 2011). A variety of factors could account for this difference, including differential manufacturing processes for bags used in Australia and the UK.

THE BIN LINER DILEMMA

Single-use plastic bags do not end their life when they come into the household from the supermarket. Table 9.2 shows the range of reuses found in a UK study. In Australia, an estimated 14 per cent of bags are returned to supermarkets for recycling (DSEWPC 2012). Australian Bureau of Statistics surveys show them to be the most commonly reused household material (89 per cent of households in 2006, up from 83 per cent in 2000) (ABS 2006). It seems that the most common mode of reuse is as rubbish bin liners. In their study, Lewis et al. (2010, 151) assumed that 19 per cent were reused in this way, 'based on an estimate of the maximum number of bin liners required by a household

Table 9.2   The reuse of lightweight carrier bags

Reuse application	Per cent
Use as a bin liner in kitchen	53
Put rubbish into it then throw it away	43
Use as a bin liner in rooms other than kitchen	26
To store things at home	14
For dog/cat/pet mess	11
Reuse for shopping other than supermarket	10
Reuse for supermarket shopping	8
For packed lunches	8
Carry other things in when going out	4
Put football/wellington boots in	1
Garden refuse	1
Give to charity shops	1
Keep bottles/cans in for recycling	1
Other uses	2

Source: Adapted after Edwards and Fry 2011, Table 4.5.

each week'. They also note that the Australian reduction in HDPE bag use between 2002 and 2007 'was offset by a small increase in consumption of "kitchen tidy" bags (bin liners)'. For example in 2006 consumption of plastic carry bags fell by 3455 tonnes, while consumption of kitchen tidy bags increased by 364 tonnes (Lewis et al. 2010, 146, citing data from Hyder Consulting 2008).

There is some disagreement between analysts over the worth of the bag to bin liner pathway. Edwards and Fry (2011) lean towards advocating reusing plastic bags as bin liners, while Hyder Consulting (2008) do not believe it outweighs the overall disadvantages of HDPE bags. But what do householders think and do? Bin liners were an active topic of discussion in our ethnography. The practice of reusing HDPE bags as bin liners was common among participants, as described by Loretta:

> I reuse the Woolies [Woolworths supermarket] shopping bags, so I never buy bin liners, but I use any plastic bag that comes into the house for that purpose, and nothing out in the bins outside, we just tie the end off and put them in.

Marie described a finely calibrated set of decisions by which she usually shopped with reusable green bags, but left these at home when she needed a new set of bin liners:

I've got a dispenser [of used supermarket bags], so it's running low now, so this weekend when I go and do my shop I won't take my own bags, which I usually do, yeah, and I'll get about ten and that'll do, that'll last me for a few months ...

Decisions about different sorts of bin liners for different parts of the house vary with the degree of messiness and disgust around the different types of waste. So for example Evette told of using proper bin liners in the main outside bin where the 'dog doo doos' went, and in the main kitchen bin:

Evette: The stuff that goes into there is the real toxic stuff, it's the gross stuff.
Interviewer: Like, you mean meat scraps and things?
Evette: Yeah, and tissues and you know, bits of scraps of things and all the plastics so, yeah, it gets rather yuck.

In other smaller bins around the house, Evette considered supermarket bags to be adequate. Only one of the households mentioned returning batches of bags to the supermarket for recycling.

The issue of 'yuckiness' raised by Evette connects to what alternative bin lining practices might be if plastic bags were banned entirely, or if households choose to avoid them completely. Composting can certainly reduce the amount of 'wet' waste going into the system, but it is difficult to remove it entirely, and some common household waste is not compostable. The most commonly recommended practice is to wrap scraps in newspaper. In fact there are several YouTube clips demonstrating newspaper origami for bin liners (see http://www.youtube.com/watch?v=BfEX85V9n8w for an example). As print newspapers become less prevalent, possibly to be replaced by electronic versions only, it is unclear how many households still have a ready supply of newsprint for this and related purposes, and for how long. A similar dilemma confronts parents of children using disposable nappies – where the temptation to wrap the entire used 'bundle' in supermarket plastic bags (or now in dedicated, smaller HDPE nappy bags) is ever-present, exacerbating risks of human waste going into kerbside waste collection. Such practices are frequently the subject of scorn from parents using reusable nappies instead (who must flush waste or dispose on-site as part of the cleaning and laundering process) (Chapter 1). They also point to the subtle agency of such seemingly inanimate things as nappies and plastic bags – tempting humans into patterns of relating despite their better judgement.

WHAT TO DO?

In focusing on supermarket plastic bags and bin liners we have only scratched the surface of the multiple issues around packaging, plastic and the household.

We have not dealt with the tear-off plastic bags typically used for fruit and vegetables, or with the overall level of packaging in shops. We have not delved in detail into biodegradable or compostable supermarket bags because they are of variable composition in different places and are still a small component of usage. The significant land use impacts associated with corn production make it unlikely however that 'corn starch plastic' can ever be considered an environmentally friendly choice. Indeed, in view of the oil intensity of corn production, producers of such bags will need to work hard to demonstrate that they are not just another form of intensive fossil fuel use. Edwards and Fry (2011, 60) found that 'starch-polyester blend bags have a higher global warming potential than conventional polymer bags, due to the increased weight of material in a bag, higher material production impacts and a higher end-of-life impact in landfill'.

Lifecycle analyses for different bag types are technical and tricky. Although different studies vary in their findings, the broad trends are clear. Getting food and other groceries into the average household requires the use of bags or other containers. The best ones are reusable bags of different sorts, but they must be used often, and over a longer period of time, to be worthwhile. Of single-use bags, the notorious HDPE supermarket bag actually comes out best in all studies we have found, partly because it is light and thin. Single-use paper bags are worst, contrary to some green consumerist understandings. The environmental performance of any bag type can be increased by reusing it as much as possible, whether for shopping or as a bin liner.

The everyday context of bag use within households can shift the assumptions built into lifecycle studies in different directions. People use, reuse and circulate bags in complex ways. Our ethnographic examples indicate the detailed way in which householders negotiate questions about mess and waste as they make decisions about different sorts of plastic bags.

Widespread horror at the prevalence of 'plastic' in modern life may be misplacing limited environmental attention away from issues of greater urgency and overall impact, for example climate change. On the other hand, plastic bags provide a good example of the ways in which everyday dilemmas can engender public passion and participation. The invasion of plastic into our lives, the profusion of bags spilling from our kitchen cupboards and the distress of marine birds and mammals all appear connected in such a way that many people want to do something about it. This is the energy that we need to harvest and intensify for the less visible problems.

10. Driving cars

In 2011 over 60 million new passenger cars were produced, around 165 000 new cars every day (OICA 2012). With the exception of 2009, due to the global financial crisis, the total number of cars produced globally each year has continued to grow. More people now than ever before are driving cars to move from place to place. But car-driving is more than moving from points 'A' to 'B'. For many people, car-driving is regarded as a right, an integral component of contemporary citizenship (Cresswell 2006). Car-driving is framed as progressive, efficient and modern, as well as offering social status, convenience, comfort, safety and freedom (see Featherstone et al. 2004).

A further complication is the role car-driving plays in a fragmented social life across extensive physical distances, requiring juggling and managing a personal timetable (Urry 2004). Cities are commonly reliant upon and designed around car mobility. Car-driving has facilitated spatial reconfiguration of urban form including suburban growth and dormitory villages, and physical separation and distancing of schools, homes, work, leisure sites and families.

Underlying urban design and infrastructure often make driving a necessity to participate in society. Furthermore, worldwide, governments are implicated in promoting the car as a way of securing capital accumulation. In comparison to the car, infrastructures for other modes of transport including walking, cycling, trains, trams and buses are underfunded. Such modes of transport are rhetorically positioned as slow, impractical, inflexible, fragmented and inconvenient (Böhm et al. 2006). Globally, governments continue to be actively involved in the expansion and subsidization of the car industry but at the same time are alert to the greenhouse gas emission of cars and diminished oil reserves. The household sustainability dilemmas of car-driving therefore evoke questions about the roles of personal transport choices, government regulation and automobile manufacturers.

SUSTAINABLE MOBILITY?

Three very different motivations drive political discussions around the sustainability of car-driving. First, it seems possible that cheap oil may run out in our lifetime (Rapier 2012). World demand for oil is outstripping supply; reflected in increasing prices at the bowser. In 2010, the world consumed more than 80 million barrels of petroleum per day; exceeding production by over 5 million barrels per day. In 2005, OPEC abandoned its price band because of the erosion of excess oil production capacity in the wake of strikes in Venezuela and military action in Iraq. Since 2005 crude oil prices have remained volatile. Prices have risen steadily since the 2008 global financial crisis, supported by rising demand in Asia, and interruptions in supply in the wake of the Libyan civil war.

Second, a substantial amount of the world's oil reserves are located in countries other than the major oil-consuming economies. Politicians in many countries are concerned with security-of-supply as 'national' oil reserves and production decline and access to overseas resources can no longer be guaranteed.

Third, cars are a major and growing source of acid rain, smog and greenhouse gas emissions. Further upstream greenhouse gas emissions come from road construction, fuel extraction, processing and manufacturing. Nevertheless, attention remains focused on tailpipe emissions from passenger cars. Globally, passenger cars account for roughly 10 per cent of carbon dioxide emissions (645 Mt CO_2-e) (ITF 2009). Carbon dioxide emission trends are projected to increase by nearly 50 per cent by 2030 and more than 80 per cent by 2050 (IEA 2009a). In no small part, higher carbon dioxide emissions from passenger cars can be attributed to the ever-increasing number of private cars. In 2010, the number of registered passenger cars topped 1 billion, increasing from 500 million in 1986 (Ward's Automotive Group 2011). Vehicles in operation equate roughly to a ratio of one vehicle to 6.75 people. China, along with India, accounted for the largest growth in vehicle registrations, increasing from around 61 to 78 million vehicles overall, and 19 to 20 million vehicles in 2010. The largest numbers of registered vehicles are in the United States (239 million), Japan (73 million), Germany (45 million), Italy (41 million), France (37 million) and the United Kingdom (35 million) (Ward's Automotive Group 2011). World Bank (2005) data suggests that countries ranked highest for number of passenger cars per hundred people are all in Western Europe, with the exception of New Zealand, Canada and Australia. Luxembourg, Iceland, Germany, Switzerland, Malta and Austria each have over 50 passenger cars per hundred people.

THE DILEMMAS OF REDUCING TAILPIPE EMISSIONS

How much carbon dioxide each car contributes is related to the distance travelled, on-road fuel economy and the energy per unit of fuel. One strand of subsequent government policy focuses on the substitution of petrol for 'greener' fuel types that are already available, such as hybrids that use electricity and petrol to power motors, liquefied petroleum gas (LPG), biofuels such as ethanol, and electric vehicles. This strand of policy intersects with questions of energy security and the dependence on imported oil in the European Union and the United States. Three core barriers are: lack of price-competitive alternative fuels; limited fuel distribution systems to petrol pumps; and inadequate supply of non-petroleum fuelled vehicles. The United States 2007 Energy Independence and Security Act mandated that consumption of ethanol would be 36 billion gallons by 2022. United States production of corn ethanol has risen from 50 million gallons in 1979 to 13 billion gallons in 2010. Likewise, in 2009 the EU adopted the Renewable Energy Directive, which included a 10 per cent target for the use of renewable energy in road transport fuels by 2020.

Hybrid cars were introduced onto the market because of their 'environmental friendliness'. The marketing pitch of the car aimed to tap into a cultural logic that driving is deeply embedded in identity, emotions, appearances and aesthetics. In the mid-2000s Toyota intensified its marketing pitched to British and European consumers. However, motoring authorities immediately questioned the 'eco-friendly' attributes of the Toyota Prius, pointing out that the diesel Volkswagen Lupo would actually do more miles per gallon. Jeremy Clarkson (2004) from the *Top Gear* television series thus mocked the Toyota Prius as 'a very expensive, very complex, not terribly green, slow, cheaply made and pointless way of moving around'. Similarly the Lexus RX400 was pitched as: 'High performance. Low emissions. Zero guilt' (Environmental Leader 2007), but the advertisement did not contain any CO_2 emissions or fuel consumption data. As noted by Mackay (2009), the Lexus RX400's carbon dioxide emissions were 192 g/km, higher than the average new car sold in the United Kingdom (168 g/km), and the European Union proposal for a target of 120 g/km by 2012.

Proponents of biofuels hope to cut carbon dioxide emissions by turning plants into liquid fuels, substituting petroleum with fuels produced from rape seed, sugar beet, corn, sugarcane and palm oil. The biofuels industry is poised for rapid growth, with mandated legislation in the United States and the European Union. However, some scientists warn against positioning biofuels as a 'fix'. First, the benefits from burning biofuels are always contingent upon previous uses of the land on which they are grown (Fargione et al. 2008). Converting rainforests, peat lands, savannahs or grasslands in Brazil, Southeast Asia and the United States creates a long term 'biofuel carbon debt' and biodiversity loss, offsetting greenhouse gas advantages from not burning fossil fuels. Second, the production of biofuels competes with food, diverting previous food crops into fuel – a critical food security issue. Hence, some scientists (e.g. Tilman et al. 2009) advocate for producing biofuels that: (1) involve feedstocks that neither compete with food crops nor directly or indirectly cause land-clearing; (2) involve growing perennial plants on degraded lands abandoned from agricultural use; (3) use crop residues such as corn stover (leaves and stalks) and straw from rice and wheat; and (4) use municipal and industrial materials rich in organic matter that are conventionally destined for landfill (see also Chapter 11 on aviation biofuels). Nevertheless, biofuel crops often deliver modest power per unit of area (Mackay 2008). It would be impossible to meet present world demand of around 80 million barrels per day, even if all of the agricultural land in the world was devoted to biofuel production.

Electric cars offer possibilities to reduce dependency on fossil fuels and carbon emissions without compromising food security. There are a growing number of advocates for the electric car (Motavalli 2011). The average running cost of the REVA electric car in London was 21 kWh per 100 km – about four times better than an average fossil fuel car (Mackay 2009). Furthermore, in carbon emissions, 21 kWh per 100 km is equivalent to 105g CO_2-e per km (assuming that electricity has a footprint of 500g CO_2 per kWh), just under half of the average carbon emissions of a new car sold in the United States (255 g/km) and nearly two-thirds of the average carbon emissions of a new car sold in the United Kingdom (168 g/km). Evidence from European car markets show that a battery range of 50 miles (80 km) is enough for most people, most of the time. Such statistics have helped encourage some of the largest automobile companies into the electric car market.

A second strand of policy encourages research into the redesign of vehicles that are aerodynamic, smaller and lighter. Fuel economy standards are not new. For example, in the United States in the wake of the 1973 oil crisis, Congress passed the legislation that led to the first fuel economy standards for passenger cars. Internationally, fuel economy standards vary between countries. In the past decade, European Union regulation has been the most proactive in reducing fuel consumption, in light of commitments under the Kyoto Protocol (Feng et al. 2007).

Fuel efficiency between models of new cars for sale varies greatly. For example in 2012, among the top ten selling new vehicles in Australia fuel efficiency varied from a Mazda 3 using roughly five litres per 100 km, to a Toyota Hilux 4x4 that uses 10.6 litres per 100 km (DITRDLG 2012). Moreover, key findings from analysis of automobile fuel economy and carbon dioxide emissions from 1970 to 2005 in the United States, Japan and Europe suggest that on-road fuel economy was offset by an upward spiral of the average annual distances cars were driven (Schipper 2005). Sprawling cities, traffic congestion and buoyant economies counteract efficiency improvements from smaller, more fuel-efficient vehicles. In the United

States, carbon dioxide emissions attributed to motor transport only fell during the 2008–2009 economic recession.

Therefore, a third strand of policy thinking focuses on changing travel demand within transportation systems. This strategy includes a mix of policies including carbon emissions taxes; changing driving behaviour (driving a slower, steady speed on long-distance travel, downsizing, car-pooling, car-sharing); and educating households with 'green' driving tips (purchasing low-carbon vehicles or using alternative forms of transport, particularly on shorter journeys). The UK Government linked automobile carbon dioxide emissions to road tax. In 2011, only owners of cars with carbon dioxide emissions less than 100 g/km were exonerated from the additional tax. Owners of cars emitting over 255 g/km of carbon dioxide were charged £455. Such market-based approaches often rely on education policies to correct market failures, and to ensure that individuals have greater information to practise more sustainable consumption. For example, the Australian Department of Infrastructure, Transport, Regional Development and Local Government (DITRDLG) (2012) invited consumers to: 'Make a smarter choice... By choosing a greener vehicle; you can make a real difference.' The United States Environmental Protection Agency's Green Vehicle Guide (EPA 2012a) education campaign proclaimed: 'No matter what size vehicle you need, you have green options!' The underlying assumption is that when consumers are equipped with the facts and figures of the impact of their consumption practices – particularly when there are clear cost savings – individuals will follow the prescribed 'rational' course of pro-environmental action.

Here we need to consider the dilemma raised by the so-called 'value-action gap' – that is the disjunction between public awareness, concern and knowledge about climate change and limited behavioural responses (Lorenzoni et al. 2007). This is particularly the case for habitual car journeys (Whitmarsh et al. 2011). Surveys consistently illustrate that despite respondents being knowledgeable and concerned about climate change, people tend not to modify their everyday driving practices. As Lakoff (2004, 14) argued, 'the facts bounce off'. If the science does not support a person's cultural values, a person's everyday goals may remain unchanged, regardless of the credibility of scientific data on climate change. This dilemma for policy makers is hardly surprising to anyone familiar with the cultural norms, habits and passions of driving a car.

CAR CULTURES

While some decisions drivers make are economic – purchasing petrol, insurance, models and brands – the values supporting and shaping the decision to drive a car are fundamentally cultural (Miller 2001). Preferences and behaviours stem from social norms, individual habits, embodied knowledge and passions. One important Australian study for instance tracked how car use was gendered: the car is used as a 'management tool' for complex daily routines and to maintain notions of 'good mothering' (Dowling 2000). Another revealed that cars formed the basis of vernacular creative pursuits in industrial cities – among working-class young people otherwise demonized as uncouth or lacking creativity (Warren and Gibson 2011). Cars meet very different (perceived or actual) needs.

While car cultures are quite distinct between countries and social groups, similarities exist between the sociotechnical and political practices that shape personal mobility around the car (termed automobility) (Edensor 2004). Automobility produces and reproduces norms of convenience, comfort, cost-effectiveness, style, progressiveness and efficiency (Featherstone et al. 2004; Urry 2004). Theresa Harada's ethnographic study of car drivers in an affluent Sydney suburb (Waitt and Harada 2012) revealed strong thoughts on the importance of branded cars. According to Phil, a general manager in his 40s:

> Obviously I am well aware of the whole impact of greenhouse gases and everything else. It is interesting at work at the moment... we are recycling waste and doing everything we can to reduce our impact on the environment. And, where you really got me thinking is, hang on, I am doing all of this and I am jumping in this two-tonner vehicle [BMW] that burns up a lot [of] fuel. ... Do I really need it, a three-litre, six-cylinder, two-tonne car? No I don't. I could get around in a Honda Jazz or something similar. But I do enjoy driving that big car.

Cars are a means to cultural capital, as Lucy (19-year-old student) explains: 'We've got big cars here. Around here people tend to keep up with the Joneses... At the moment four-wheel drivers are popular, but if hybrid cars became popular then people around here would drive them'.

Social hierarchies pattern different forms of mobility. Particular models and brands of cars are viewed above others, and private cars are seen as above walking, cycling or public transport. The seemingly rational choice of driving a less-polluting car is trumped by the pleasures and social status of driving a particular model and brand of car. Jim (retired manager, aged over 60) understood the car as a basic necessity of affluent suburban dwelling: 'I suppose the car for me is just a necessary mode of transport. You just think about it as one of the necessities of life. I know people use public transport but you wouldn't want to live out here if you wanted public transport.' Everyday mobility choices are a mechanism of individual and social differentiation. Driving a private car not only helps to navigate between neighbourhoods, home and work when doing everyday mundane tasks, but also sustains a sense of self.

Cars differentiate places through the multiple synchronized routes, routines and rhythms of driving. Edensor (2010, 5) argues that rhythms of mobility flow 'contribute to the spatio-temporal character of place'. In Harada's ethnography, rhythms of car mobility intersected with discourse of freedom, safety, convenience and efficiency:

> Public transport would not get me to where I need to go during the day; within the timeframes I need to operate... It is just not a practical proposition (Phil, general manager, aged 40–50).
>
> If it [public transport] worked into my schedule with what I do, when I was working. You are so busy, you do things for convenience (Anna, teacher, aged 40–50).
>
> I have missed trains by minutes and have to wait half an hour for the next one, and that really frustrates me because when you have a car you can go when you want to. You are not stuck to timetables (Lucy, student, aged under 20).
>
> For someone of my age [using public transport] would be a ridiculous hardship (Harry, retired engineer, aged over 60).

Car journeys are seen to maximize use of clock-time, enabling access to destination choices, facilitating personal rhythms of movement. Car journeys are anticipated as flexible, eliminating the need to become part of a mobile public constrained by tickets, collective scheduling, connections and timetables. Despite the ever-present risk of traffic jams, the rhythms of car journeys are cherished as economical, smooth, seamless and instantaneous.

Driving practices are also unconscious, everyday sensual engagements. Sheller (2004, 288) argued that driving has 'transformed the way we sense the world and the capacities of human bodies to interact with the world through the visual, aural, olfactory, interceptive and proprioceptive senses. We not only feel the car, but we feel through the car and with the car'. Bodily sensations of car mobility sustain particular conscious and unconscious connections with driving. Norman (2004) revealed how automobile manufacturers are very aware of the importance of bodily sensations in the design of cars. As Phil's experience illustrates, purchasing decisions about a new car may involve a sensual appreciation:

> Like I said to you, I did test drive the hybrid Lexus ... it is really boring. The whole layout of the car. I didn't really like the driving position in it. I just felt completely disconnected from the driving experience, it was all very smooth, and it was almost like getting into an automatically driven bus or something. The nice thing about the BMW is that they still put sports suspension on it, you can feel the road a little bit more and the steering is very precise. You can have the performance of the engine if you want it to be a quite racy vehicle.

Several ethnography participants spoke of the importance of the bodily sensations of driving comfort:

I suppose you feel slightly unsafe on the train – you see some people get on the train who worry you. That's the detraction of getting a train I think – you feel slightly unsafe. Driving your own car you feel safe and comfortable... We are all spoilt now. It is impossible to live without a car. At our age we are looking for comfort and safety and everything. We want our car! Our cars! (Patrick, retired tradesman, over 60).

Light drizzle; what a comfort it is to go door to door (almost) by car, low heat on, listening to 2RN [ed note: an Australian radio station]. Warm, quiet and comfortable. With the car windows closed, it almost excludes the rest of the world ... I'm beginning to think I DO have a love affair with my car (Harry, diary entry, retired, over 60).

Sensations of comfort encourage the driver to dwell in an individualized, cocooned and privatized space that facilitates a distancing from the stress and dangers of the world beyond the windscreen (Sheller 2004). People are temporarily 'cocooned' from intrusive sounds, smells, temperature and sounds beyond the car – enveloped in the illusion of safety.

WHAT TO DO?

Driving cars is an unambiguously unsustainable practice. Yet encouraging people not to purchase a car or to drive less is far from convincing. Purchasing a car has become a major item of individual consumption that provides social status and differentiation to its owner. Car culture sustains the 'good life'. Car mobility is also entangled within finely tuned rhythms of modern parenting (Dowling 2000) and has itself reorganized the temporal and spatial rhythms, routes and routines of everyday life. Alongside increased choices of where to live, work and play, car ownership has patterned social life, as people find themselves juggling aspects of their lives over increased distances. Caring for family and friends may require driving thousands of kilometres each year. It is therefore naïve to assume that people are going to give up driving.

One starting point is to think carefully about how to reduce the total distance a car travels in a year. Car-pooling and sharing cars are options to lower regular commutes to work or school-runs. Through sharing a car there are possibilities for the apparent flexibility and freedom of automobility, while minimizing new production and the number of cars on the road. Sharing cars nevertheless conflicts with intense privacy values (Klocker et al. 2012) – a direct parallel to televisions (see Chapter 13).

Walking or cycling over short distances reduces CO_2 emissions given the fuel inefficiencies of journeys less than 8 km (5 miles). Walking and cycling cultures require provision of facilities and road layouts that favour pedestrians and cyclists rather than cars. Beyond the scope of the household, effective mechanisms to reduce car use include congestion charges and improving public transport. Another issue for households considering purchasing a 'new' car is to remain alert to the 'green claims' of automobile companies and consider downsizing their vehicles. Through downsizing come possibilities of the lower carbon emissions of a smaller, lighter vehicle, while allowing continued expression of preference and taste. Government-sponsored 'green vehicle guides' provide comparative data on fuel efficiency, air pollution and carbon dioxide emissions. Nevertheless, when the car remains an expression of individual status, style and emotional attachment, whether 'green' credentials will be central to most 'new' car purchase decisions is moot. Legislative measures beyond those of the household therefore remain imperative, including mandates on fuel efficiency and carbon emissions on new models.

More difficult to unravel are the frictions associated with our cultural norms of seamless time. We increasingly come to expect seamless time in the way we juggle our days. But of course in practice seamless time is already confounded in lots of ways: we resent waiting ten minutes for a late train, but not ten minutes filling the car with petrol. Key is perhaps a sense of movement and agency (something smartphones (Chapter 14) arguably address by 'making productive' waiting time (Bissell 2007) on public transport systems). Such perceptions of time are not innate, but might prove difficult to budge.

11. Flying

Personal and business travel has exploded in recent decades, underpinned by accessible and cheap air travel. The development of rapid and affordable air transport is also inextricably linked to the structure of the global economy. The global aviation industry has fostered aeromobility – a 'complex set of social representations, imaginations and practices as much as the outcome of technological advances', in which travel is 'associated with all number of bodily pleasures and excitements – as well as anxieties' (Adey et al. 2007, 776) (see also Chapter 10). These anxieties may well include the fear of losing such mobility if the debates, conflict, and rationalizations surrounding air travel and climate change are anything to go by. Travellers fear the loss of freedom and the return of constrained worlds, governments fear loss of economic activity, and the industry fears loss of its *raison d'être* and continued profits. Long the focus of battles with residents over airport development and expansion, the aviation industry is now squarely in the sights of those who argue for significant changes to current lifestyles to cope with climate change. In part this is an issue for an affluent and largely Western elite who account for a disproportionate proportion of carbon emissions from air travel (Gossling et al. 2009; Wilkerson et al. 2010). Through its global economic role, the fate of the aviation industry will also affect many, beyond affluent travellers.

Aviation growth has been facilitated by developments such as the introduction of wide-bodied jumbo jets in the 1970s. The scale of the industry can be seen in some basic figures: in 2006 the global commercial aircraft fleet flew 31.26 million flights, burned 188.20 million metric tonnes of fuel, and covered 38.68 billion kilometres (Wilkerson et al. 2010). Since the 1970s world air travel has grown at an average annual rate of 5 per cent, approximately twice the annual growth in global gross domestic product. In North America, revenue in passenger kilometres (one paying passenger transported one km) has increased about seven times between 1971 and 2005 (Belobaba and Odoni 2009). Growth is expected to continue, and even under conservative projections will nearly double in the next 10 to 15 years (OECD 2012). A key driver of growth has been liberalization of the aviation industry which, among other things, has facilitated the success of low cost carriers (Fu et al. 2010). As a result, in the United States the real median price of a fare fell 40 per cent between 1980 and 2005 (Fu et al. 2010). The US Department of Transportation

(DOT 2008) estimates that just over half of all air travel is for 'pleasure/leisure', more than for business (41 per cent) or 'personal business' (8 per cent).

SUSTAINABILITY ISSUES

A whole host of sustainability issues accompanies flying. These include the impacts of airports such as noise, air, and water pollution (Black et al. 2007; Suau-Sanchez et al. 2011) as well as in-flight issues such as waste (Li et al. 2003). This chapter focuses on two key issues driving change and debate in and around aviation: fuel use and emissions.

Aviation fuel use corresponds to 2–3 per cent of the total fossil fuels used worldwide and 80 per cent of aviation fuel is used by civil aviation (Penner et al. 1999). At the moment fuel is supplied largely from fossil sources, a source that will decline in coming decades. Technological change, largely operational management and aircraft engineering, reduced fuel consumption by 70 per cent in the four decades to 2000 (Marias and Waitz 2009) and from eight to five litres per passenger km in the last two decades (Zhang et al. 2010). While fuel efficiency could improve by up to 50 per cent by 2050, it is most likely that fuel efficiency in 2050 will be 'largely determined by existing technologies' (OECD 2012, 47) and fuel use may well double with industry growth by this time (CSIRO 2011). This is partly due to technological limits but also due to the complexity of the industry and costly fleet turnover (Bows et al. 2005).

In an attempt to secure fuel supplies and lower the aviation industry's carbon emissions, airlines are actively developing biofuels (see also Chapter 10), which have been used in fuel mixes for passenger flights since 2011(IATA 2011). Claiming that aviation biofuels can reduce emissions by up to 80 per cent, the industry expects them to play a key role in meeting their carbon reduction goals (IATA 2011). Lifecycle analysis demonstrates that this may be possible, although 80 per cent is at the upper end of possible savings, and such figures do not include emissions associated with potential land use change (CSIRO 2011).

The production of biofuels is not without its critics – such as the Aviation Environment Federation (see for example, Gazzard 2008). As discussed in Chapter 10 in relation to cars, biofuel production can drive land use change – such as clearing of carbon sinks including forests – thereby increasing emissions. While land use change is complex, biofuel production has been part of overall increased demand and competition for land and other resources used for food and bio-energy production (Harvey and Pilgrim 2011). This changes production patterns, the consequences of which have included (and may continue to include) displacement of food production, expansion of production into habitats that are carbon sinks, and land-grabs (Harvey and Pilgrim 2011;

van der Horst and Vermeylen 2011). While the industry argues that such problems will be avoided in so-called 'second generation' feedstocks that do not necessarily require good quality land (IATA 2011), such optimism is not universally shared. For example, van der Horst and Vermeylen (2011) argue that it is the very marginal lands that the industry envisages being used for biofuels which are often most important for marginal peoples, and which are vulnerable to acquisition by commercial interests.

Closely linked to aviation fuel use are emissions during flight. These include carbon dioxide, water vapour, nitrogen oxides (the emissions of which tend to increase with rising fuel efficiency), condensation trails, soot emissions, and sulphate emissions (Penner et al. 1999). It is possible that aviation emissions at high altitude are particularly potent but there is uncertainty around this. More generally, the aviation industry makes a disproportionately large contribution to global emissions; air transport accounts for about 1 per cent of the world's gross domestic product but is estimated to contribute about 2–4 per cent of global greenhouse gas emissions (Zhang et al. 2010). Future scenarios parallel projected growth rates in civil aviation. For example, a meta-analysis of 30 scenarios concluded that aviation CO_2 emissions will increase faster than the overall global economy (Gudmundsson and Anger 2012).

For households, one flight can counter any other efforts to be more sustainable. Wright et al. (2009) describe an Australian commuter who saved 3000 kilograms of CO_2 per year by riding 10 kilometres to work instead of driving. According to the Qantas online carbon offset calculator (Qantas 2012b), if he took one return flight from Sydney to Los Angeles his emissions would be 3362 kilograms of CO_2. If he flew premium economy, business or first class, his carbon emissions will be between one and a half to about three times those of economy (Kollmuss and Lane 2008; ICAO 2010).

Alongside fuel use, improvements in emissions are possible and expected. Fuel efficiencies on current planes are now comparable to modern European cars (OECD 2012) and, *excluding* possible potent high altitude impacts of aviation emissions, planes on average produce slightly less emissions than cars (Wright et al. 2009). Table 11.1 breaks this down somewhat for the UK, highlighting the low emissions of coach and rail and the high emissions of short air trips (for which land transport is an alternative). Improvements in aviation notwithstanding, the sheer volume of kilometres travelled in planes and the expected growth in air travel means that aviation emissions are likely to grow in coming decades (Marias and Waitz 2009). Even under a technically 'optimistic' scenario, 'demand management or a reduction in the current rate of aviation growth through other means is required if the sector's emissions in 2030 are to reduce below their 1990 level' (Bows et al. 2009, 34). Clearly such strategies are unlikely to find favour in the aviation industry. Moreover, they run counter to national government efforts to facilitate growth in and via

Table 11.1   Selected UK emissions per passenger km, 2011

Transport mode	CO_2 emissions per passenger km
Average car	151.0g (average 1.6 passengers)
Average motorcycle	139.8g
Average coach	36.4g
National rail	65.1g
Domestic aviation (between UK airports)	195.2g
Short haul aviation (typically within Europe, to 3700 km)	114.7g
Long haul aviation (non-European destinations, over 3700 km)	132.0g

Note: Generalized figures – details can be found at Department for Environment, Food and Rural Affairs (2011). See Wilkerson et al. (2010) for a slightly different view on aviation emissions.

Source: Adapted from Department for Transport (DFT n.d.).

airport expansion and construction, in itself an investment in carbon-intensive infrastructure (Bows 2010). In Australia, the debate over a second airport in Sydney illustrates this commitment. At a 2011 press conference the Federal Minister for Infrastructure and Transport said:

> unless there is a second airport…[the result] is delays for passengers, inconvenience and a cost to economic growth and jobs… Sydney is Australia's global city. It is the city that global citizens want to fly into more so than any other airport around Australia. And it's vital that that infrastructure be kept up to date. That is why we need a second airport. (Albanese 2011)

Government support for aviation is further evident in low carbon tax imposts on passengers. For example, to offset the cost of the Australian carbon pricing scheme, introduced in July 2012, Qantas will be charging domestic Australian passengers AU$1.82–AU$6.86; it also charges AU$7 for international return flights from Australia to Europe to cover the cost of the EU Emissions Trading Scheme (Qantas 2012a).

Clearly, air travel is an activity that has a major environmental impact and the extent to which it is successful in dealing with issues such as emissions will have a global significance. It would therefore seem a likely target for reduction, for an individual or household wanting to live more sustainably. Yet people fly for a wide range of (to them) perfectly good reasons and relinquishing aeromobility will be profoundly challenging. Flying also supports many economies such as those dependent on current models of tourism. There are multiple influences on the decision to fly and difficulties in carbon-offsetting flights.

For many, particularly in the West, aeromobility has gone from a luxury for the wealthy to an affordable and normalized part of life in the space of a few decades (Adey et al. 2007). The tourism industry seems to agree. In 2002, Eugenio Yunis, the World Tourism Organization's director of sustainable development, defended people's 'insatiable appetite' to fly the world:

> The important point is people's desire for mobility. The expansion of tourism demand is something that cannot be controlled – we live in a free society and the solution is not to stop people travelling or price them out of the sky with taxes. So we have to make these trips as least damaging as possible. (Walters 2002)

Certainly, the mobility brought by air travel fulfils many social and economic functions, such as facilitating business and family relationships (so-called 'love miles', Monbiot 2007). Nevertheless, our survey and ethnographic data did reveal uneven patterns of propensity to fly: the largest segment of survey respondents (48 per cent) had not flown at all in the past 12 months. Just over 20 per cent, the next largest segment, had flown only once. Smaller proportions had flown between five and ten times that year. One 45–54-year-old male had flown 25 times in that year (typical of a fly-in, fly-out mine worker – a peculiarly carbon-intensive labour market phenomenon servicing Australian and Canadian remote resource regions; see McIntosh 2012). Notwithstanding the democratization of travel, socio-economic disparities still profoundly structure access to air travel.

The issues above raise the bigger question: to what extent is unconstrained mobility a feasible goal for society? Given increasing capacity for telecommuting and operating business meetings using videoconferencing technologies, how do travellers dodge this question? Their answers illustrate the significance of the social and cultural issues that keep people flying.

Lassen (2010) considered the rationalities that underpinned Danish 'knowledge' workers' decisions to fly. Even in this 'dematerialized' sector, work-related flying remained important and, despite high levels of environmental concern and practice at home, employees continued to fly for work. They did so for several reasons including 'culturally embedded expectations' and judgements of success in their organizations (Lassen 2010, 741). These include expectations that academics present their research at international conferences. (We can attest how difficult it is to budge this: at one prominent international conference in the UK we sought to present via Skype from Australia, presenting enormous logistical difficulties for organizers, and crippling time differences for us, presenting in the middle of the night.) In addition to such drivers, Lassen (2010) identified several individual factors: the role of work-related air travel in supporting a cosmopolitan identity, and the opportunities flying provided for professional development, an escape from busy lives, and for holidays.

Similar tendencies among leisure travellers indicate that environmental concerns about air travel alone will not curb demand. Cohen et al. (2011) argued that contemporary demand for leisure-related air travel was akin to 'binge flying', an addiction to carbon-intensive mobility, participation in which we embrace even though we know it to be bad for us (cf. Davidson 2012). Air travel mobility has within a few short decades become associated with an ideal of freedom (attained via holidays based around air travel) that will be difficult to disentangle (Cohen et al. 2011). Air travellers aware of the impact of flying express 'sentiments of guilt, suppression and denial of air travel's climate impact' (Cohen et al. 2011, 1085). Such sentiments are not, however, enough to prevent people from flying for leisure. For example, Cohen and Higham (2011, 334) found that long-haul flights to New Zealand were justified by travellers as extraordinary 'once in a lifetime' trips. Such travellers are often willing to trust in technical solutions such as carbon taxes (Barr et al. 2010; Barr and Prillwitz 2012).

For those who have decided to fly, the main option available to mitigate the impact of flying is to purchase voluntary carbon offsets (VCOs). Rather than reduce emissions at source (by not flying), by purchasing a VCO a passenger can pay for reductions in CO_2 elsewhere (Broderick 2009). This is usually an option at the time of ticket purchase and the amounts are modest. At the time of writing in May 2012, Qantas quoted an offset price of AU\$12.82 per person for a flight from Sydney to Los Angeles and Virgin Atlantic quoted £7.06 for a flight from London to New York. In theory a passenger can achieve carbon neutrality for their flight if the quantity offset is equal to the amount of per capita emissions generated by their flight. Examples on airline websites of projects funded by VCO payments include renewable energy schemes, tree planting to sequester CO_2, and distribution of fuel-efficient stoves in the global south.

Are VCOs effective? Pearce (2007, np), writing in *New Scientist* claimed 'the answer is a resounding maybe'. Highlighting the potential role of VCOs in justifying business as usual, this ambivalence was echoed in a recent blog posting:

> Yup folks, that's right. I spent \$34 (Canadian) on, umm … What exactly??? This is the problem with offsets. I always end up feeling a bit, well, conned. My three choices for offsets that Air Canada's partner Zerofootprint provided were planting trees, recycling tires and capturing landfill gas for use in generating electricity. But shouldn't we be doing these things anyway? More to the point, wouldn't we be doing these things anyway? (LaForce 2010)

Unlike regulated 'cap and trade' systems where a central authority such as a national government distributes tradable permits with an upper limit on total emissions, VCOs exist in a largely unregulated, uncapped global trading system (Broderick 2009) that struggles to ensure that they meet the criteria necessary for credibility (Kollmuss et al. 2010). This is particularly the case where VCOs come from activities that wouldn't happen in the absence of an offsets incentive (for example see Hogarth 2008; Schmidt 2009). While VCOs may have the potential of 'co-creating' environmental value whilst reducing the 'environmental loads' (Gossling et al. 2009, 17), they have been criticized on a variety of grounds. These include: low participation by passengers; variable calculations of the offsets required; the imposition of afforestation on developing countries and their potential to justify the acceptability and growth of air travel; and to act as a disincentive for structural change in the consumption of energy (Gossling et al. 2007; Mair 2011). Gossling et al. (2007) argued that participation in VCOs would need to increase by a factor of 400 over 2006 levels to reduce aviation emissions by 10 per cent and that they are an 'ambiguous' and 'risky' tool with which to tackle the impact of air travel.

WHAT TO DO?

It is clear that flying for work, leisure or family reasons will continue to have major environmental consequences. It is likely that people will continue to fly despite environmental concerns, that governments are unlikely to take action to curb flying (they are more likely to encourage it), and that airlines will pursue a modified business as usual approach. Relying on the citizen-consumer to mitigate the impacts of flying has its limits (Barr and Prillwitz 2012). That said, if choosing to fly, you can reduce the payload of the aircraft by taking minimal luggage. Paying for a VCO is also a step to take, even if one has doubts about them. These are often offered through the booking process. There is however nothing to stop a traveller purchasing their offset elsewhere, after consulting independent organizations such as Carbon Offset Watch and Carbon Offset Research and Education. The only other option is to not fly, joining the many without the means to choose to fly, or at least reduce one's flying. Some, such as George Monbiot (2007), argue that this is necessary. It is worth remembering too that regular air travel has become normalized only since the 1990s. Before this, flying was for many a once in a lifetime occasion. Assuming one is *personally* willing to forgo the mobility afforded by flying, this potentially raises difficulties in both workplace and family realms, where the pressure to fly to conduct business or visit elderly parents is likely to remain. Geographical quirks such as the fly-in, fly-out phenomenon to mine sites make this especially so. This decision is ultimately not only an individual one, or at least not an easy one for most individuals, and groundwork may need to be done to develop alternatives, change workplace expectations, or prepare one's family or employer. This may include using options such as videoconferencing (for which in turn there are both positive and more muted assessments, see Lassen et al. 2006; Arslan et al. 2011). If flying is out, there are, of course, other modes of transport such as train and bus. These are likely to be particularly suitable for the large number of air travellers who fly relatively infrequently and domestically (Gossling et al. 2009).

In a work setting, such modes may require changes to expectations about the relationship between travel and other tasks and responsibilities. Mobile technologies may facilitate such changes, allowing for example work to continue on a train. In a model where same-day intercity business travel is common, such changes will have far-reaching consequences. For other travel, it may require allowing more time for travel on family visits or, in the case of holidays, a different approach to travel. In this respect the 'slow travel' movement offers an alternative to much contemporary tourism. Slow travel is diverse, but is associated with less travel-intensive tourism, use of transport such as train, coach, or bike, depth of experience, and transport being part of the trip rather than just a means of getting to a destination (Dickinson et al. 2011).

Should such an approach catch hold, it will change the geography of tourism and travel and be accompanied by a reassessment of how we perceive and value time. This is a big ask, yet this model does not request that mobility be relinquished but that it be reappraised and its outcomes be reassessed, and that alternative, not quite obsolete, mobilities be (re)imagined. In this way, recently gained yet apparently taken for granted aeromobility might be seen for what it is: merely one form of mobility among other possibilities, and one that is an artefact of our times and expectations, and therefore open to change.

12. The refrigerator

Refrigerators keep our drinks cool and our meat, fruit and vegetables 'fresh'. They extend the length of time that perishable food items like meat and dairy products can be stored by holding off the constant decay of the world. Mechanical domestic refrigeration has transformed routine food preservation, shopping, cooking and eating. Fridges allow us to bulk-buy, organize food in shelves and bins, prepare meals in advance, serve chilled desserts and perfect the art of serving up left-over meals. Alongside artificial cooling to preserve food, the thing that enabled the entry of the refrigerator into the kitchen was its labour-saving potential, performing the work of organizing food.

In countries like Australia, Britain, Canada, New Zealand and the United States the ordinariness of the refrigerator in domestic kitchens has only become so since the 1960s. The first mechanical refrigerators for domestic rather than industrial use, like those manufactured by Kelvinator and its rival Frigidaire, were manufactured in the United States during the early 1900s. Cowan (1985) outlines the rise of the electric-powered compression refrigerator, over gas refrigeration during the 1920s in the North American market. Whereas there was only one gas refrigerator company in the United States at this time (Servel), there were several highly capitalized corporations seeking market and profit through the design and manufacture of the electric-powered compression refrigerator, including Westinghouse, General Electric and General Motors. Fridges as we know them today originated in this first great period of the electrification of the home, when an entirely new suite of domestic technologies, including home electrical cabling itself, patterned the economy (Mitchell 2008). Nevertheless, as Nye (1990) observed, for some time the refrigerator was considered a luxury and novelty item. Today's ubiquity only arose in the post-war boom.

Nowadays refrigerators are commonly regarded as a household 'necessity' rather than 'luxury' item. They have become incorporated into household routines and help sustain understandings of freshness. Chilled or frozen foods within their 'best before' dates are normally understood as 'fresh', rather than something just harvested or killed (Isenstadt 1998). Like many other everyday household objects in this book (Chapters 6, 8, 15), we may only think about the refrigerator when it draws attention to itself by breaking down or needing replacement (Verbeek 2004).

SUSTAINABILITY ISSUES

What are the sustainability issues arising from the always on cooling and organizing of food in refrigerators? Early debate centred on chlorofluorocarbons (CFCs). CFCs are a synthetic refrigerant, developed in the laboratories of General Motors' Frigidaire Division in the 1930s, amid intense competition for market share. CFCs were a substitute for potentially explosive and toxic refrigerants such as sulphur dioxide, ammonia and methyl chloride. Donaldson et al. (1994) argued that domestic refrigeration could not have become widespread without the non-flammable, odour-free chemical attributes of CFCs (primarily before Freon-12). Believed to be 'safe', it was seemingly unimportant that CFCs imperceptibly leaked on a daily basis from domestic refrigerators. Molina and Rowland's (1974) benchmark paper in *Nature* linked CFC emissions with ozone depletion. In 1979, the World Meteorological Organization (WMO) and the United Nations Environment Program (UNEP) organized the first international meetings to address the issue of ozone depletion. In 1987, the Montreal Protocol was ratified, to end CFC production by the end of the century and minimize the depletion of the ozone layer. The UNEP (2007) reported that the number of new domestic refrigerators containing CFCs had fallen from over 63 million units in 1992 to around 3 million units in 2004.

Carbon dioxide emissions were a second focus of debate, this time in relation to climate change (IEA 2003; UNEP 2007). Refrigerators contribute to greenhouse gas emissions in two ways. First is an indirect contribution from the energy required to operate the vapour compression cycle. Refrigerators consume electricity 24 hours a day, seven days a week, over a product life-cycle of 20 years and beyond. This indirect contribution is related to the number, volume, energy efficiency and everyday practices associated with the refrigerator. In OECD countries almost every household has a refrigerator (98 per cent). In Australia, not only do most households have a refrigerator (99.8 per cent), but one-third (34 per cent) have two or more in use (ABS 2008a). In 2000, there were an estimated 391 million domestic refrigerators in OECD countries, an increase from 315 million in 1991. Worldwide there are an estimated 1.2–1.5 billion domestic refrigerators (UNEP 2007). While the number and capacity of refrigerators in OECD countries increased, total electrical consumption of the OECD stock of refrigeration appliances is estimated to have stabilized by 2000, at around 314.6 TWh (terawatt-hours) annually. As an OECD country average, keeping food cold accounts for 13.4 per cent of the total electricity consumption and is therefore an important source of carbon dioxide emissions (UNEP 2007). According to the International Energy Agency (IEA 2003), in the year 2000 the average household in OECD Europe consumed 700 kWh/year of electricity for refrigeration, compared with

1034 kWh/year in Japan, 1216 kWh/year in Australasia, and 1294 kWh/year in North America. Fridges are certainly one of the biggest power consumers within households (West 2011).

Second, domestic refrigerators contribute to climate change from hydrofluorocarbon (HFC) refrigerants emissions into the atmosphere during use and end of life. HFCs were the most common replacement for ozone-depleting CFCs in domestic refrigerators. However, HFCs are known to have a high global warming potential (GWP) (Johnson 1998). The Kyoto Protocol recommended the reduction of six categories of greenhouse gases including HFCs. Alternative refrigerants are being studied, including carbon dioxide and hydrocarbons (HC) such as propane, butane and isobutene (Wen-Tien 2005; Sattar et al. 2007; Mohanraj et al. 2008). Such studies outline the pros and cons of so-called 'natural refrigerants' as a substitute for HFCs. For example CO_2 as a refrigerant requires high pressure, and has lower energy-efficiency under high ambient temperatures. In contrast, HCs have been shown to work as a refrigerant in domestic refrigerators (Hammad and Alsaad 1999), but pose explosive and exposure risks.

TRANSFORMATIONS AND COMPLEXITIES

Efforts to improve the sustainability of refrigerators must also remain mindful of the different ways in which this domestic appliance is incorporated into the everyday social, political and economic ordering and scheduling of household life, as well as the provisioning of food (see Shevchenko 2002; Perez 2012). Technologies of refrigeration brought not only a control over temperature but possibilities to transform household practices and domestic spaces.

Domestic refrigeration helped accomplish new kitchen practices, making previous practices and technologies of food preservation almost redundant (Watkins 2008). Passive cooling technologies such as cellars, cool-rooms, pantry windows and marble shelves of larders have largely disappeared from the design of contemporary Western houses. Equally, certain food preservation practices that do not require a constant energy source are no longer as widespread, including sand pits, smoking, salting, jamming, home-bottling and pickling (Madison 2007). The domestic refrigerator animates the landscape of household sustainability in other ways too: influencing not only what we eat and our shopping patterns, but also how much food we buy (and waste).

While social commentators and politicians may lament the loss of community and increasing social isolation of our cities, this is rarely connected to how the refrigerator has changed how people shop. In our household ethnography one of the most frequent satisfactions of refrigeration was the practice of food-provisioning through bulk buying, visiting supermarkets on a weekly basis rather than small local grocery stores on a daily basis. Stockpiling was never spoken about as a practice against the possible disappearance of foodstuffs. Instead, in a world of perceived food abundance, the refrigerator was a strategic weapon against price and for managing time. George spoke about buying in large quantities as a way of balancing 'need' and thrift:

> I live by myself, but I try to buy enough of frozen stuff. I don't eat much meat at all. I don't eat red meat as a rule, I buy chicken and stuff, looking for it when it's on special. But I buy it sort of not in bulk but in larger quantities that I don't immediately need and freeze stuff.

Amanda, who lives on a disability pension with her partner, spoke of bulk buying as a practice that fitted in accordance with her household's environmental ideals and thrifty aspirations:

> The reason we got a big one [refrigerator] was because we just weren't fitting [our food] into the small one. We don't like using lots of poisons around the house so one of the ways to reduce cockroaches and ants and things was to have a fridge that was big enough to take everything and lots of jars and things to put other stuff in. So that was a sort of a conscious decision and so that we could get frozen organic meat and store it rather than just buy it locally which tends to be very expensive.

Typically, participants spoke of bulk buying as a coping mechanism to sustain 'normal' family life. Emily lives with her husband and is a working mother of five children. In describing her use of the refrigerator Emily said:

> Well, buying in bulk, I buy milk in bulk. Probably 12 litres at a time and just put it in the fridge. It saves me having to go every day to the shop.

Refrigeration has facilitated changes in food-provisioning, making possible the weekly or fortnightly supermarket shop. That bulk purchasing has become 'normal' also reveals changes in food supply systems and perceived food abundance. Slowing down the decay of foods by bacteria has facilitated the global 'reach' of dairy, fruit and vegetable produce. Households have come to expect a regular abundance of foodstuffs sourced globally to refill their refrigerators on a weekly or fortnightly basis. Restocking a refrigerator takes for granted the networks of global transport systems and the agricultural and food processing practices leading to it.

GENDERED DIVISION OF LABOUR AND REFRIGERATION

Practices associated with refrigeration are gendered, with sustainability implications. Given the gendered division of household labour, mechanical refrigeration impacted on the lives of women more than most men (Watkins

2008). The social reproduction of home relied upon women's designated domestic labour and responsibilities in food handling, food preparation and family health. How the mechanical refrigerator first came to belong in the kitchen was tied to not only general economic growth, mass production factory systems, electricity grids, suburban living, and women's entry into the paid labour force, but also the gendered norms of who was responsible for feeding and shopping.

Home, gender and technology are co-produced through their use (Wajcman 1991; Silva 2000). Electric appliances become domesticated and normalized through everyday household labour practices of home-life (Morley 2003). The entry of the refrigerator into the domestic kitchen is an integral consequence of the story of domestic hygiene in the nineteenth century (Robinson 1997). Following Louis Pasteur's discovery of bacteria, by the late 1860s and 1870s household 'germs' were transformed into a scientific 'fact'. Marketing refrigerators in the 1920s and 1930s relied upon the fear of germs in the spread of disease within the home (Tomes 1998). In the 1930s, buying a white, streamlined, steel refrigerator and keeping it clean in the kitchen mirrored back white American middle-class cultural standards of thrift, hygiene, proficiency and modern homemaking (Nickles 2002). As women increasingly entered the labour force the popularity of refrigerators, like that of the freezer, was entwined with manufacturers' promises of alleviating the difficulty of juggling domestic and paid work (Yeandle 1984; Shove and Southerton 2000).

In our ethnographic work it was primarily women who spoke about how the refrigerator enabled them to better organize time through practices of cleaning, cooking and restocking. Tracy, who lives with her husband and two teenage sons, praised the anticipated 'convenience' of a frost-free refrigerator in helping her juggle her part-time job, voluntary work, domestic housework and leisure activities. Likewise, Janine, a part-time working mother of two primary school aged children, talked about how refrigerators provided the 'convenience' of ready access to 'basics' and taking advantage of leftovers. Notwithstanding women's greater involvement in paid labour, women always reported spending more hours conducting housework than men (cf. Baxter et al. 2003). Sustainability implications arise because it appears that women in affluent households are more likely to restock, clean, dispose of 'waste-foods', look after and make replacement decisions about fridges.

LIVING WITH REFRIGERATORS

The energy efficiency of refrigerators is central to regulations designed to lower greenhouse gas emissions. In the United States federal legislation passed in 1987 established minimum energy performance standards for refrigerators that took effect in 1990. As a consequence, the energy efficiency of refrigerators for sale in the North American market has steadily improved. Taking 1972 as their baseline year, Von Weissäcker et al. (1997) traced how the power consumption of the average model sold in the United States fell from 3.36 kWh/year to 1.52 kWh/year in 1990, and to 1.16 kWh/year in 1993 when Federal Standards were tightened. Gradually, the most inefficient refrigerators are being phased out.

An Energy Star labelling scheme first introduced in the United States and adopted by many OECD countries now facilitates comparison of energy efficiency between models at the point-of-purchase (see: http://www.energystar.gov/). Energy ratings are displayed on refrigerators to help consumers compare the energy consumption. Worldwide, there are around 60 countries that have mandatory comparative energy labelling, including member states of the European Union (EU Energy Label) and Australia (see: http://www.energyrating.gov.au; Winward et al. 1998).

Results from a 2008 survey conducted in Australia suggested that householders ranked energy star rating as more important than price or brand when buying a new refrigerator (ABS 2008b). In contrast, results from a qualitative research project conducted with self-identified 'green' consumers (that is people most likely to buy an energy-efficient white good) in Britain suggested that participants were more likely to compromise on environmental and ethical criteria on the purchase of white goods than small electrical appliances (McDonald et al. 2009). Consumers may rank energy star rating over price or brand as their ideal, but in practice these may be compromised by availability, price and brand. Moreover, as McDonald et al. (2009) found, while some 'green' consumers had considered doing without certain small electrical appliances, none had considering living without a fridge. Within a period of 50 years living with a fridge has become completely normalized even among a group that might be considered the most 'environmentally conscious'.

Alongside 'eco-labelling', the decision to purchase an energy-efficient refrigerator is being encouraged by the introduction of government subsidies for replacement, combined with information campaigns. In Australia, a 'Fridge Buyback' scheme was introduced under the New South Wales Government's Energy Saving Scheme. Households were offered free collection and AU$35 rebate for a pre-1996 refrigerator (see http://www.fridgebuyback.com.au/index.html). Such schemes, however, rub up against other norms encountered throughout this book (from clothes to furniture, televisions to mobile phones); of accommodating things that retain use value. Not all things are thrown away after regular use, but are stored, repurposed, accumulated (Gregson et al. 2007). Fridges are one of those things.

Table 12.1   Number and age of refrigerators per household Illawara region

Number of operating fridges	Frequency	Number manufactured before 1996
One	769 (54.5%)	169 (22 %)
Two	544 (38%)	190 (35%)
Three	97 (6%)	34 (35%)
Four	10 (1%)	10 (100%)
Five or more	4 (0.5%)	4 (100%)

Source: Author survey results, 'Tough Times? Green Times? A survey of the issues important to households', 2009.

Strandbakken's (2009, 149) work on Norwegian households confirmed this, suggesting that 'consumers are reluctant to exchange products that work'. Instead, Strandbakken (2009, 149) traced the 'afterlife' of energy-inefficient refrigerators as they were sold, given to charity, family or friends or relocated in second homes, garages and basements. New energy-efficient fridges added to the total population of refrigerators, rather than simply substituting for inefficient ones. Results from our household survey confirmed Strandbakken's argument. Some 45 per cent of households had more than two operating refrigerators (Table 12.1). Crucially, a significant proportion of second fridges were older, energy-inefficient models built before 1996.

Arnold and Anne described how they replaced their fridge without hesitation when they redesigned their kitchen to suit their new taste (see also Chapter 8). Arnold and Anne are retired, and their children have grown up and left home. Rather than disposing of the old refrigerator their second refrigerator was accommodated into the garage to suit routines of sports and entertaining family and friends at home:

> Arnold: We had a new kitchen put in and it [the refrigerator] wouldn't fit in the new kitchen so we had it down…
> Anne: … stairs as a drink fridge… It's very hard to sort of dispose of something when it still works. So we do use them both…
> Arnold: People would come and say, gee, I love your beer fridge.
> Anne: So then when we came down here that fridge fitted in that spot [in the garage] so then the one that we had upstairs that's the one that's out there. The fridge mechanic said whatever you do, if you've got to get rid of a fridge get rid of that one [the newer fridge] because he said the old one will keep going forever and a day and the new ones won't.

There are two points to draw out here. The first regards links between appliances and home design. Anne and Arnold are not alone in reconfiguring the kitchen space. Kitchen makeovers are commonplace in affluent consumer societies (Freeman 2004). The kitchen is no longer a hidden away site of food-preparation chores, but has been re-imagined as the 'social hub' of the 'good family life' (Johnson and Lloyd 2004). Kitchen renewals usefully retrofit homes to suit shifting social norms and needs, but are also a means to taste and social distinction (Southerton 2001). Beyond the impacts galvanized through replacing cabinets (Chapter 8), kitchen makeovers work against sustainability by propelling some individuals towards the second fridge. The second – and arguably more knotty – issue is the tension between getting rid of something that works, to replace it with something more energy-efficient. In Anne and Arnold's case, the mechanic was the key gatekeeper adjudicating on whether the old appliance should be retained, irrespective of its energy use burden.

The second-hand market was a second pathway to an afterlife for old energy-inefficient refrigerators. Some participants justified only ever owning second-hand refrigerators in terms of waste reduction, reducing the number of products shipped around the world and supporting smaller businesses rather than large multinational corporations. One participant, John, was reluctant to purchase a new fridge because possession of such a new object would not align with his own environmental ideals and aspirations.

However, for the case of refrigerators, the relationship between 'alternative' supply chains and sustainability is not perfectly aligned. While the second-hand fridge minimizes the minerals and energy used in the manufacturing process, the sustainability of domestic refrigeration must always consider the implications of constantly operating the appliance. Buying a second-hand refrigerator without an energy rating, or sharing older refrigerators with family and friends, may work against efforts to improve household sustainability offered by more energy-efficient refrigerators built after 1996.

Purchasing a new fridge also felt wrong in many households balancing 'need' against thrift. Mary, retired, who lives alone after her children moved out of home, wrangled 'need' and thrift, in her approach to replacing her refrigerator:

> My fridge, I had to give it up after 30 years. I had bought it in 1973 and it was still working. It was huge and it was the old gas, you know, whatever at the back. I just exchanged it for a smaller one for myself. I've got that plastic crap thing now... That's your world, it's all throwaway.

Disappointment expressed about the absence of an old refrigerator reveals changing ideas about domestic appliances. The plastic rather than metal components of modern domestic appliances were understood as material evidence of poor quality and planned obsolescence. Among older generations in particular this created added resistance to purchasing a new energy-efficient model. For households balancing need and income, purchasing a new refrigerator is not quite right, particularly when the old appliance is still work-ing. The absence of a new model is interpreted optimistically as the value of making-do with an older model.

WHAT TO DO?

One option is to consider living without a fridge. If this is too radical many organizations and energy utilities have started to provide households with information on energy saving using behaviour, including operating tempera-ture, installation, position and maintenance. Some models are generally more efficient (such as those with top-mounted freezers and without ice-dispensers), and dimensions and volume are further factors. What is a 'safe' food storage temperature seems also to be a contested issue. Other recommendations include: observing the correct operating setting for the ambient temperature (refrigerators are classified in one or more of four climatic classes), prevent-ing placing hot food in the refrigerator, thawing frozen foods in the refriger-ator and keeping the refrigerator full to reduce the frequency that the compressor switches off and on (Geppert and Stamminger 2010).

How to best dispose of an 'old' refrigerator is contingent on the size and age of the appliance as well as where you live. Refrigerators manufactured before 1993 will normally contain CFCs in both the refrigerant and the insu-lation. By law, the CFCs usually must be removed and recovered before the metal can be recycled. Refrigerators containing CFCs can no longer be refur-bished and resold. This brought an end to the once legal international trade in second-hand fridges to West Africa and Eastern Europe (HoC 2002). Some retailers, refurbishment businesses and scrap metal recyclers distance them-selves from helping consumers dispose of old fridges. In several countries the state has stepped into the void, offering cash rebates.

The more difficult dilemma is exactly when to replace a refrigerator. As shown in this chapter, the 'rational' choice to replace an old, inefficient fridge with a more energy-efficient new one collides with other values – such as a preference for well-made, lasting appliances. The point of traction is also the point of friction.

13. Screens

The television, arguably more than any other domestic technology, influences how we 'design our spaces, habits and even emotions' (Lanvin 1990, 85). After sleeping and working, it is the activity that most occupies daily time, the 'single most common form of domestic recreation' (Bennett et al. 1999, 67). There is a television device now for every four human beings on earth, making it second only to the mobile phone (Chapter 14), 'one of the most popular pieces of electrical and electronic equipment in our society' (Hischier and Baudin 2010, 428). Television is a powerful symbol and presence in everyday life: an archetypal technology of consumerism, a sign of escape from peasantry, an icon of suburban life, a principal means of communication and imagined community. This chapter accordingly discusses the sustainability dilemmas of televisions in the household.

It is difficult to separate discussion of the environmental sustainability impacts of television from that technology's myriad other impacts. Television influences all kinds of things. It informs and entertains, seduces, and pacifies; television is a means to advertise excessive consumption to broader populations, and can lead to extreme human sedentarism (Hu et al. 2003). Television watching has been shown to reduce teenagers' consumption of healthy fruit and vegetables, contributing to growing problems of child and youth obesity (Boynton-Jarrett et al. 2003). Beyond the immediate impacts associated with the production and running of televisions are broader entanglements and practices in the home.

What exactly constitutes screen-based entertainment has also changed since the advent of the television, as personal computers, laptops, eReaders and tablets proliferate. For this reason, the chapter also encompasses related discussions of these screen technologies.

SUSTAINABILITY ISSUES

Estimation of the actual contribution of TVs, computers and other screen devices to overall household energy use varies. The International Energy Agency estimated that electronic devices as a group account for 15 per cent of household electricity consumption (IEA 2009b), and rising fast. So too is overall global television ownership, which has been above 95 per cent of households in all OECD countries for some time. In China, television ownership has expanded rapidly and is now above 80 per cent; in India 30–40 per cent (though rapidly growing) (World Bank 2012). Computers are also increasingly ubiquitous. In Australia, 83 per cent of all households had computers in 2011, up from 70 per cent five years previously (ABS 2011a). Over half of all UK homes now own laptop computers, up from 30 per cent in 2006 (Spinney et al. in press). With ownership of televisions, computers and tablets in highly populous countries such as India and China growing quickly, the International Energy Agency predicts that the energy consumed by consumer electronics goods will triple globally by 2030 (IEA 2009b).

Modelling of the total environmental impact of such devices is limited. Hischier and Baudin (2010) conducted lifecycle assessment of plasma, liquid crystal display (LCD) and cathode ray tube (CRT) televisions – their conclusion was that all three had similar patterns of high environmental impacts in production and use (compared with distribution and end-of-use). Of the three, plasma screens appeared to have the lowest impact per screen inch. Nevertheless, televisions vary greatly. Plasma screen televisions are generally larger, requiring more overall energy than smaller LCD and CRT televisions (up to four times as much, according to Hischier and Baudin (2010)); hence nearly 600 kg CO_2-e (or 6000 bags of CO_2) are emitted to run a large flat screen (plasma) TV annually, while watching a small 50 cm screen television averaged just under 100 kg CO_2-e emissions annually (Wright et al. 2009).

Watching practices also vary greatly, so that the 'average' use may not reflect the diversity of how most households watch television. Watching television for one hour per day contributes an average of 80 kg CO_2-e annually (Berners-Lee 2010), but in many households, televisions are left on or on standby throughout the day. Resulting 'leaky' energy use and emissions multiply substantially. According to Hischier and Baudin (2010), 'active' standby (where a device typically 'updates' in standby, such as with programme guides) can use up to twenty times the electricity of 'passive' standby modes – depending on the television set in question. Energy use studies by Australian consumer group Choice demonstrated that standby and on-power energy use in digital set-top boxes could vary as much as 400 per cent across different brands, all within the same price bracket (Choice Australia 2009).

Augmenting the picture are national energy systems and regulatory frameworks. According to Hischier and Baudin (2010) what contributes most to impact during the use phase of televisions is the production mix used for the electricity consumed. If coal-based electricity is used, then the impact of using a television over its lifetime can be double that of its initial production. In some countries, clear energy labelling on televisions is required, or due to be phased in, while in others consumers are left in the dark. At the time of writing, on the website of the most prominent national home electronics retailer in Australia it is possible to obtain 'specs' on a variety of its wide range of televisions on sale (dimensions, screen definition and resolution, audio systems and connectivity), but not the power rating, watts or energy usage.

Because televisions and computer equipment are complex, high value-added technical objects, they require significant energy and chemical inputs. Berners-Lee (2010) estimated that manufacture of a 42-inch plasma screen produced 220 kg CO_2-e. Williams et al. (2002, 5504) estimated that a single 2 gram 32MB computer chip needed 1.7 kg of secondary fossil fuel and chemical inputs in production – its materials intensity 'orders of magnitude higher than that of "traditional" goods'. Screen devices are also increasingly made in complex and often opaque global production systems that link raw materials, design, prototyping, automated parts manufacture, assembly line production, marketing and distribution. With these come problematic environmental and social impacts dispersed across continents and contexts: demands for both common and rare earth minerals; energy and chemical inputs for manufacture; waste products; labour exploitation and transport fuel emissions (see also Chapter 14). Tracing the inputs and production impacts of the hundreds – if not thousands – of components that are assembled together to make televisions, computers and other screen devices is a monumental task.

OBSOLESCENCE

In wealthy countries, the cultural norm is increasingly to replace televisions and computer equipment because of rapid product advances, rather than because of the failure of existing products. The resulting e-waste is simply massive. According to Ongondo et al. (2011) each year 20–50 million tonnes of waste electrical and electronic equipment is discarded globally. In the EU, televisions, DVD players, laptops and phones were the second-most common form of e-waste (after large household appliances such as fridges (Chapter 12) and washing machines (Chapter 7)). Britons discard in the order of 100 million electronic items annually (Ongondo et al. 2011). In the United States, an estimated 300 million electronic items were removed from households, with two-thirds still in working order (EPA 2008). Barely a quarter were recycled; the bulk left in storage, unused, within the home, or disposed in landfill. In Australia, e-waste is growing at over three times the rate of general municipal waste (ABS 2006). E-waste is also exported from rich countries (over 6800 tonnes annually from Germany alone – Ongondo et al. 2011) to India, China, Nigeria and Eastern Europe, some for remanufacture and refurbishment, but others destined for landfill, where downstream environmental consequences are felt. By and large, consumers rather than manufacturers remain legally responsible for e-waste (Lepawsky 2012).

Making matters worse is built-in obsolescence. In Hischier and Baudin's (2010) study the assumption was that the lifespan of a plasma screen television was eight years. In the 1960s, the lifespan of a computer was approximately ten years; by 1998 this had dropped to 4.3 years; now less than two years. According to the National Recycling Coalition (2007) between 1997 and 2007 nearly 500 million personal computers became obsolete in the United States.

Recent years have witnessed a further acceleration of replacement cultures for televisions, fuelled by analogue-to-digital replacement, plasma and LCD screen improvements (high definition, LED, HDMI inputs, USB connectivity, surround-sound capability, built-in Wi-Fi), and the much-advertised introduction of 3D televisions. Such tactics are well known within the electronics, television and entertainment industries as strategies to continue to extract profits out of platform replacement (Christophers 2009). The motivation is sustained profit and viability for electronics and entertainment firms, but the result is continued expansion of the physical production (and environmental and social impacts) of such goods.

Even where households might wish to reduce the environmental burden of their consumption by purchasing electronics goods made to last, such items may be difficult to judge at face value, at point of purchase. One brand might have a preferential reputation over another, but whether this genuinely reflects quality, or rather brand and price-point positioning, is moot. Dedicated homework is required from consumers – point-of-purchase marketability instead rests on technological features, picture or sound quality. A further dilemma is that information necessary to assess total environmental impact of new screen products such as laptops, tablets and eReaders is complicated and costly to accrue, and is inevitably redundant upon publication – the products themselves being rapidly superseded by newer models and methods of manufacture in the time that it takes to fully investigate and model total environmental impacts.

CONVERGENT SCREEN CULTURES

For four decades television technologies were relatively stable (cathode ray tube) and grew steadily in numbers, while computers were separate (albeit also rapidly proliferating) items. Nowadays households watch increasing amounts of content via computer screens, laptops, tablets and even smartphones (Chapter 14). In the United States television ownership numbers actually dropped in 2011 for the first time in two decades (Stelter 2011), in part a function of recession, but also from young people leaving home and moving to college and relying on online and smartphone viewing instead of purchasing television sets. According to one industry report, 'the percentage of consumers watching broadcast or cable TV shows, movies or videos on TV, in a typical week, plummeted from 71 percent in 2009 to 48 percent in 2011' while '33 percent of consumers now watch shows, movies or videos on their PCs, and 10 percent are watching such programs on their smartphones' (Accenture 2012). According to Spinney et al. (2012): 'computing' is no longer a reliable term to describe a discrete set of activities using computers – and likewise, 'television' itself is increasingly an antiquated term for a physical technology superseded by computers.

Flat-screen, touch-screen, storage and wireless technologies thus reconfigure a range of practices: viewing, working, downloading, playing. As yet, however, there has been little engagement with the household energy implications of changing the use of computers and other screen devices (Spinney et al. in press). A BBC report on the total environmental impact of television – from programme production through distribution, device production and home viewing – concluded that the most substantial contributing stage to energy use was in viewing, but this varied enormously depending on the device used: 76 per cent of the total for terrestrial television, 78 per cent on desktop computers, and as low as 37 per cent on laptop computers (BBC 2011). The trend towards viewing on smartphones, laptops and tablets means that a greater proportion of programming could likely be accessed on lower-power devices.

One of the promises of convergence is elimination of the need for multiple appliances: a tablet or laptop fulfils work needs and acts as an entertainment tool later in the day, while also rendering obsolete a host of other separate gadgets and appliances and 'content' commodities (CDs, DVDs, books). In the same vein are thin- and cloud-computing – where several small devices without hard drives or moving parts are used by individuals, interacting with only a single server/storage device located elsewhere in the house, or remotely (CanyonSnow Consulting 2009). The potential exists for overall material consumption (if not energy consumption) to be reduced through lower overall need for separate technologies to watch, read, listen, work and access information.

Adjudicating such promises against actual consumption levels is, however, difficult. In hypothetical terms it might be feasible to compute the accumulated lifecycle impacts (energy, water, carbon and so on) of a range of separate appliances and goods, versus a single, multifunctional, convergent screen object such as a laptop or a tablet. This was the goal of Kozak and Keoleian's (2003) industry-funded (AT&T) comparison of the lifecycle assessments for an eReader versus 40 printed text books. Their conclusion was that eReaders required less energy than to produce the paper needed for the stock of books, and that they also produced lower CO_2-e emissions. But that assumes that individuals purchase said items at more or less the same rate. In reality, people buy books, CDs and DVDs, and watch, read or listen online or via computers at vastly different rates. Lifecycle assessments by their definition model averages, but in the case of entertainment appliances and content, wild variations from the norm *is* the norm (Rainey 2012).

As well as the longevity of the television, DVD player, laptop or tablet, the durability of 'content' platforms varies enormously. With books, once the paper is made and books are produced, the carbon is captured and there are no significant further emissions impacts – although accumulation of books requires more bookshelves and larger homes (see Chapter 8). Lifespan can potentially be infinite, given the right storage conditions. There is likely to be an ongoing market for antique books (although second-hand books may fall in value appreciably), while exactly how long electronic files of books 'last' is bound up in more complicated questions of the durability of the eReader or computer equipment, and of the intellectual property rights pertaining to downloaded files – both unknown quantities. In our household ethnographies, some were much more reluctant to throw out books than other items, although this caused heated arguments at times over storing and stockpiling:

> Will: There's boxes of books and stuff that you had …
> Michelle: Yeah, I know. I can't throw books out. That's difficult for me.
> Will: They were your father's …

Meanwhile, CDs and DVDs, once made, purchased and stored, only require the electricity needed to power the hi-fi or home entertainment system to be enjoyed – minimal amounts in the whole scheme of things, compared with heating, cooling and cleaning (Chapters 5, 7), yet their lifespan too is unclear. In the 1980s, then new CDs were marketed positively as being able to last 25 years, a length of time that seemed considerable, but which might trouble people now that their CD collections are approaching that age. Cassette tapes have not lasted at all well; vinyl has lasted much longer (and as objects might well outlive CDs). All this has to be somehow balanced against the gains made in consumption of digital files rather than physical platforms: eReader users purchase fewer printed publications, and users of digital music, television and movie files no longer need to accumulate bulky piles of discs, or cupboards or shelves to store them in. Ultimately, it is impossible to know exactly which options for screen technology are better. As Rainey (2012, 1) describes, 'It is not just about numbers, such as tonnes of CO_2, raw materials and waste, but also about human behaviour and interpretation of the impacts.'

EVERYDAY SCREEN PRACTICES

Environmental impacts of various screen technologies also need to be understood within varying – and sometimes contradictory – domestic contexts. In the case of telecommuters, home computer use may be intensive, but is necessary to earn an income. Households bear the burden of such energy use to perform a job. In other cases, intensive energy use for screens amplifies already existing socio-economic inequalities. In Australia, the lowest household income groups are much less likely to have computers in the home (40 per cent without computers) compared to the wealthiest (2 per cent without computers): thus the rich are contributing to higher levels of emissions *and* benefiting from computer use professionally and personally (ABS 2011a). In our survey, the poorest households were also the least likely to own LCD or plasma screen televisions.

Sustainability dilemmas are refracted through factors of household size, family relationships and sharing practices. Whether one laptop or tablet serving both work and entertainment purposes outweighs separate TVs and computers and piles of CDs, DVDs and books is contextual within families and everyday behaviours. It also relates to the materiality of the things concerned. A family with several members all watching a single television together is likely to be less resource-intensive than each using separate screen appliances. The latter is more likely, however, if tablets and laptops are used to watch replayed television shows, listen to music and to read books. Five family members can each draw from their bookshelf collection of shared paper books, and read in the same room together; they cannot read those same separate books simultaneously on the same eReader or tablet. In a related way, collective watching and listening also contributes to sociality and conviviality within households – a means to negotiate preferences, take turns and to talk through conflicts (Belk 2010).

Prospects for sharing (as opposed to individualized consumption via screens) are in turn a function of overall household size, urban form and internal dynamics. The trend in the global north is towards smaller household sizes on average, more single-person and couple households. This trend has been identified as a 'sleeping', but potentially catastrophic, underlying demographic shift worsening resource extraction, energy, water and carbon consumption (Keilman 2003). A smaller average household size internationally means less capacity to share appliances within homes – and greater overall demand for their production, proliferation and replication. The flipside to this is that compact urban form in combination with large numbers of small households makes possible other kinds of sharing of appliances (for example, washing machines) and entertainment content (DVD rental shops, for example, in areas where small home size absolutely precludes accumulation of bought DVDs). The small, but growing movement to cohousing and shared living schemes is another alternative (Jarvis 2011; Klocker et al. 2012). If more people on average cohabit together under the same roof, then more possibilities emerge for shared consumption of televisions, computers, CDs, DVDs and books.

Nevertheless, where larger household formations have emerged, even then everyday practices of screen watching are complicated, enabling some sharing practices while curtailing others. Within our ethnographic projects were ten participating extended family households (defined as one family plus at least one other relative) (Klocker et al. 2012). Examining their material practices in the home, as a distinct subset of the overall population, sheds light on some of these complexities (Klocker et al. 2012). Extended families tended to live together and share some appliances and domestic work (fridges, cooking duties, leftover food, and washing machines) but not others (bathrooms, cars). Extended family households were more likely to share appliances and work collectively when totally 'living together', with shared living spaces and kitchens; others by contrast were more accurately amalgams of sub-household family units, 'living together but apart' (for example, grandparents living on a separate floor of the house, or in an adapted unit separate from the 'main' family house). The latter were more likely to duplicate appliances and perform some work duties separately.

Across all ten extended family households, there was universal reluctance to share televisions. Television viewing was atomized, with an average of one television per adult household member. The shift from collective television viewing to fragmented and individualized 'multiple-screen households' has been documented elsewhere and appears common within nuclear households as well (Nansen et al. 2011, 695). In large families especially, televisions were spread across living spaces, bedrooms and kitchens. Personal televisions functioned as a retreat from 'crowded' communal areas and maintained privacy:

> I think that is a crucial thing when it comes to your personal space, is having that thing you can watch where you can switch your mind off from everything that's happening' (Mick).

Personal televisions created private 'space' and accommodated diverse viewing habits, while (additional) televisions in communal areas created opportunities to come together as a family around common viewing interests – but only when desired. For our interviewees, televisions were 'much more than a source of entertainment: this simple appliance fulfilled deeply held cultural values for both privacy and family bonding, which trumped the economic and energy savings made possible by purchasing and viewing fewer televisions' (Klocker et al. 2012, 13).

Complicating all of this further is the entanglement of screen devices in the rhythms and spaces of domesticity. What matters is not just the total stock of 'stuff' in the house, but the intricacies of appliance use and behaviour in the shifting spaces and moments of household life. As Spinney et al. (2012, 2629) argue, 'energy intensive practices such as always-on-ness, and changing computer ecologies and infrastructures, are intimately bound up with the reproduction of particular domestic imaginaries of family and home'. Lally (2002) noted that the placement of the computer within the home acted as a 'symbolic indicator' of its valued status as a household object. Some families replicate computers to allow elements of personal freedom; others buy additional desktop computers and place them in family rooms and kitchens to enable shared viewing practices, or for surveillance of children's computer and social media use.

WHAT TO DO?

Boundaries between work and home have become blurred as people read books and newspapers over breakfast, on tablets, watch missed television shows on their smartphones or even check email in the middle of interrupted nights while breastfeeding (Gregg 2011). Wireless technologies in the home have made convergence possible, enveloping the home in 'always-on' Internet and altering the coding of domestic space. Røpke et al. (2010) even go so far as to suggest that the 'always-on' configuration of wireless information technologies constitutes a new, and profound, 'second electrification' of the home, which parallels that of the early twentieth century, thus radically transforming home life, but also ramping up the spaces of the home within which electrical energy is used. Every part of the home is now a possible home office, but also a potential place to watch screens for entertainment. The result is that 'in most households the wireless-laptop assemblage may well be using more energy than the wired desktop assemblage that it is replacing or working alongside' (Spinney et al. 2012, 2632).

The Internet envelope does not end there. In 2008 the number of non-human things connected to the Internet surpassed the number of humans who were connected, creating a vision of the future in which every material thing is connected via Wi-Fi, screens, GPS chips and radio frequency identification tags (as many as 100 billion things, according to some estimates; see Glance 2011) – and in consequence, drawing on electrical power. We interact with screens not only for their functional purposes, but for their affective power, their symbolic meanings, with profound implications for uses and layouts of domestic space (Silverstone and Hirsch 1992) and hence for sustainability. The ultimate contradiction is that to best make use of purchased screen technologies they may need to be used intensively, which demands always-on-ness and amplifies energy use. At some point, too, rare metals used in construction of computers will run out – meaning that old devices will need to be 'mined' or used for longer before obsolescence (see next chapter). So long as we are still able to police the boundary between work and home (one ICT are rendering more porous), we would do well to share more second-hand books, go for a walk, or find other ways to be entertained and to enjoy family life.

14. Mobile phones

Mobile phones have become the world's most ubiquitous electronic product. They are rapidly increasing in use, especially in developing countries which account for the bulk of subscriptions (Ongondo and Williams 2011). According to the *New York Times* it took 20 years to sell the first billion phones, four years to sell the second billion, and two years to sell the third billion (Corbett 2008). In 2011, global penetration of mobile phones was 87 per cent in the global north and 79 per cent in the global south (ITU 2011) and 5 billion of the world's 6.5 billion people were mobile phone users. Total users are expected to reach 6 billion during 2013 (Zadok and Puustinen 2010). The first part of the twenty-first century could thus be characterized as 'the era of mobile communication' (Katz 2011, xi).

There are many benefits to mobile phone use, not least of which has been the improved, relatively low cost, and flexible communications, including in developing countries. Such benefits, however, have come with environmental costs associated with manufacturing, using and disposing of mobile phones. With the growing number of users and increasing rates of phone replacement, the number of 'retired' phones has increased faster than other forms of e-waste (Geyer and Blass 2010), despite a thriving second-hand market in the global south and profitable reuse and recycling markets.

SUSTAINABILITY ISSUES

Mobile phones are part of increasing consumer and business demand for ubiquitous and wireless access to communication services. The manufacture and use of phones is only one part of the networked communication system, which also includes the manufacture and operation of the radio access network (e.g. antenna towers and associated electronic equipment and power consumption) and the establishment and operation of data centres. The operation of the networks is, and will continue to be, one of the dominant elements of the footprint of mobile communication, especially as data-intensive uses increase relative to voice traffic (Fehske et al. 2011). Given that lifecycle and greenhouse gas emissions impacts of electronic device manufacturing were covered in the previous chapter, this chapter focuses on the use and consumption of phones themselves (but see Fehske et al. 2011). While phone manufacturing and network operation efficiencies are predicted to increase, the environmental impact of mobile phones is expected to increase for some years (Frey et al. 2006; Fehske et al. 2011). This is due to growth in the numbers of users, increasing phone turnover rates, and the growing proportion of smartphones, which have about twice the carbon footprint of ordinary mobile phones: 30 kg CO_2-e for manufacturing and 7 kWh/year for operation versus 18 kg CO_2-e and 2 kWh/year (Fehske et al. 2011). Scaled up to the planetary level, the implications are gigantic. Fehske et al.'s (2011, 59) predictions were that:

> The overall [global] carbon footprint of mobile communications increases almost linearly until 2020 with annual increase of 11 Mto [megatonnes] CO_2-e, an increase equivalent to the annual emissions of the whole of Luxembourg or 2.5 million EU households. The emissions in 2020 amount to more than 235 Mto CO_2-e, which corresponds to more than one third of the present annual emissions of the entire United Kingdom.

For mobile phones alone, Frey et al. (2006) calculated a 2001 global ecological footprint of 2.4 million gha compared to 1.7 million gha in 1999, an increase that they argued outweighed any increases in efficiencies in phone manufacture and operation. Moreover their lifecycle analysis suggested that the environmental footprint breakeven point for phone replacement was between seven and 12 years depending on assumptions regarding increased efficiencies in phone production and operation. This period is consistent with the accepted lifespan of a mobile phone which is around ten years, but is far more than the actual average replacement time which is estimated to be one to two years (Geyer and Blass 2010). Such figures include calculations for the extraction and processing of raw materials used in phone manufacture – processes (for some of the metals required) that have been implicated in illegal and socially and environmentally damaging mining operations in countries such as the Congo and China (Folger 2011). US non-government organization Earthworks estimated that 220 pounds of waste is produced to provide the gold for each mobile phone (Berry and Goodman 2006). Further insight into the scale of mineral use can be seen in a 2005 estimate by the US Geological Survey, that described half a billion used mobile phones sitting in desk drawers in the United States, comprising more than US$300 million worth of gold, palladium, silver, copper and platinum (Mooallem 2008).

The inclusion of these metals makes mobile phones one of the most valuable forms of e-waste, and recovering them a global industry (characterized by mainstream firms as well as a plethora of informal trade networks and small and 'backyard' operators). However, despite this and the fact that recycling can usefully recover 65–80 per cent of the material content of phones (Berry and Goodman 2006), and generate energy savings (Environmental Protection

Agency (EPA) 2012b), recycling rates generally remain low (Silveira and Chang 2010; Ongondo et al. 2011). Furthermore due to significant growth in mobile phone use in the global south and the international trade in used phones, a large proportion of phones are dismantled in countries without capacity for adequate and safe management of e-waste (Osibanjo and Nnorom 2008; Manomaivibool 2009). In such countries, much phone recycling is carried out informally and often by children. Phone components are burnt, processed in cooking implements, and bathed in acid to recover the more valuable substances. This is a significant problem as mobile phones can contain over 40 elements (Ongondo and Williams 2011), 12 of which are potentially hazardous (Wu et al. 2008). The majority of concerns come from the presence of heavy metals such as lead, nickel, chromium and copper (Geyer and Blass 2010). While the use of hazardous substances in phone manufacturing is declining, persistent bioaccumulative toxins (PBTs) are frequently cited as dangerous substances in mobile phones. PBTs are generally found in the printed circuit boards and liquid crystal displays of mobile phones and include antimony, arsenic, beryllium, cadmium, copper, lead, nickel, and zinc (Fishbein 2002; Most 2003; Silveira and Chang 2010). PBTs linger in the environment without degrading (Fishbein 2002). If humans are exposed to PBTs they can cause cancer and reproductive, neurological and developmental problems (Most 2003).

TRANSFORMATIONS AND COMPLEXITIES

As with televisions and computers (Chapter 13), high levels of phone replacement drive manufacturing and purchasing of new phones. Mobile phones have become the most frequently replaced electronic device in the world (Zadok and Puustinen 2010). It is estimated that over 140 million are discarded annually in the United States (Wilhelm 2012), and 18 million handsets are replaced every year in the United Kingdom (Ongondo and Williams 2011). In 1991, mobile phones were replaced approximately every three years. Typically, many mobile phone users in the global north replace their phones every 12 to 18 months, even though 90 per cent of phones are still functional when replaced (Canning 2006; Geyer and Blass 2010). How can we make sense of this? Two ways are to consider contextual drivers of phone replacement and the role of mobile phones in people's lives and identities.

While much attention and research literature around phones and sustainability focuses on materials and end of life management, Huang and Truong (2008, 330) argue that 'design may not be driving waste and replacement to the extent to which it is popularly perceived and may be a secondary factor in the problem of cell phone sustainability'. They draw attention to the paradigm of disposable technology that envelops mobile phones and other consumer electronics (Chapter 13), and especially to how a phone's perceived life is decoupled from its functional life. In our ethnography some participants inhabited this grey area between perceived lifespan and actual functionality. For Janelle: 'They're only designed to last until your contract's up and then they, and then, you know, they break and they're not under warranty any more.'

While researchers (for example, Katz and Sugiyama 2006; Ongondo and Williams 2011; Wilhelm 2012) have identified fashion, style, status, desire for new technology and features, phone damage, and battery life as factors in phone replacement, Huang and Truong (2008) draw a distinction between the reasons for deciding to replace a phone and the reasons for then selecting a particular phone once the decision to replace has been made. They conclude that factors such as aesthetics, design, technical features and functionality all motivated the selection of a certain model, but the primary reason for replacing a phone among their US and Canadian respondents was contract renewal and the offer of a new free phone if they renew a service contract. Consistent with other North American and UK research (Ongondo and Williams 2011; Wilhelm et al. 2011) they argued that the perceived lifetime of a mobile phone was most heavily influenced by contract length and renewal, a moment at which many companies offer new phones, or at least discounts on new phones, to customers. The experience of Marie, one of our ethnography participants, was typical: 'I was on a $29 plan and they just rang me up and they said we can upgrade you to this newer phone and it's got this brilliant 10 megapixel camera ... That's even better than our digital camera. I thought, OK, yeah, I'll have it'.

Somewhat surprisingly, Huang and Truong (2008) found that although their participants almost always exercised the option of getting a new phone with a contract renewal, their disposition was more complex than mere interest or joy upon receiving a new model. Participants also commonly expressed attitudes of reluctance and apathy towards the idea of replacing their phone before its functional lifetime had ended. Participants were aware of the sustainability issues involved in phone purchase and disposal. For example, one participant told interviewers of receiving a new phone at no cost: 'I don't know what of the inner workings is different, but it seems a little wasteful to me to have to get an entirely new device that is identical to the one I had before, to my eyes' (Huang and Truong 2008, 327).

Building on these findings, Huang et al. (2009) then investigated the role of national context with a study of end-of-life practices in Germany, Japan, the United States and Canada. They framed the study through what they called 'situated sustainability' – the idea that the sustainability of a device is dependent not only upon design and composition, but also on how the local environment supports certain consumption practices. They found that high transferability (the ability to transfer device ownership to others) can effectively foster sustainable practices by prolonging the device life, reducing e-waste, and limiting the manufacture of new devices and extraction of raw materials. There were high levels of transferability in Germany because phones were generally not locked to service providers, and prepaid SIM cards were readily available, inexpensive and commonly used. This contrasted heavily with the Japanese, US and Canadian contexts, in which phones were typically locked to particular providers and primarily acquired as part of new contracts. Such factors constrained the ability of users to pass on old phones to others. Additionally, participants in Japan were more data-intensive in their use of mobile phones, and thus often chose to keep rather than recycle their phones due to concerns about data privacy.

Outside slowing purchase, the main options are reuse and recycling. Recycling schemes are available in many countries and, where governments have introduced systems that require and/or generate funding for recycling, such as in Switzerland, Japan, and California, phone recycling rates have risen (Silveira and Chang 2010). However, in the absence of such systems, and depending on the disposal scheme used, once a phone is put into the e-waste stream, the nature of the international trade in e-waste may provide little guarantee that it will be reused or recycled safely and usefully (for example Kahhat et al. 2008; Mooallem 2008). Even where recycling schemes are in place and simply require consumers to drop phones off or post them to recyclers in free envelopes, Huang et al. (2009) found that people often did not take such actions, that they were unaware of the schemes or that they simply did not get around to it due to perceived inconvenience (Huang and Truong 2008; Huang et al. 2009). The default and easy action for many was to simply put the phone in a drawer and leave it there. This stockpiling is an important factor in low mobile phone recycling rates. In Australia around 40 per cent of people have two or more phones stored at home, with an estimated 19 million phones being stored in Australian homes in 2010/11 (AMTA 2011). Stockpiling occurs because throwing the small devices in a drawer and forgetting about them is easy; because phones increasingly contain data that people are concerned about and they may be unsure as to how to ensure data removal; and because people want a spare or backup phone for themselves, or to lend to friends or visitors. In this sense, use value continues to reside in a phone, even if it is only a potential value that is rarely realized. Huang and Truong (2008) suggest that there is an 'asymmetry' to the value attributed to old phones. Often, when their participants would offer their old phones to family or friends, potential recipients either did not want them, or viewed them only as a temporary solution. Further, they found that although many participants stored phones for their perceived value to others, not a single participant in the study was actually using a used phone they had received from someone else. In our ethnographic study some people maintained two phones – Jane for example had her own, older mobile phone 'for my private stuff', and a newer smartphone for purely work purposes. Management of new and old phones can thus mediate boundaries between private and work identities.

While the ongoing use value of relegated phones perhaps stands in paradoxical relationship with the rate of phone replacement, research into the social roles of mobile phones provides more consistent support for this valuation. Vincent (2006, 40) suggested that the strong emotional attachment people have to their phones results in a perception that they are 'too precious to let go'. Emotional attachments to phones are due to their role as repositories for memories, sentiments and social connections in the pictures, messages and phone numbers that they store and facilitate, the reassurance and convenience that they provide, and their engagement with all our bodily senses – they become 'icons' for who we are at the time of ownership of a particular phone (Katz and Sugiyama 2006; Vincent 2006). Vincent highlights the multiple roles of mobile phones and how they are involved in almost every change that occurs in people's everyday lives. Vincent also notes how the strong emotional attachment to mobile phones plays a role in making them too precious to let go, resulting in stockpiling and preventing reuse by others. Nevertheless, given the evidence on phone replacement, this attachment does not extend to preventing the acquisition of new phones. Our ethnography participants also discussed the agency of new phones themselves in relation to old phones – the capacities of smartphones for instance to 'invade' their lives (Gregg 2011), command attention and seduce new owners into higher usage rates (with commensurate sustainability implications). Hence Marie described how

> a couple of months ago I just upgraded. I used to have an old crappy phone. There it is, that old red one. I've just upgraded to a smartphone and now I'm finding myself looking at it a lot more, whereas before with that other one I'd very rarely use it. But now that I've got this smartphone where I've got access to the Internet I find myself, oh, I wonder what's in the *Sydney Morning Herald*? Oh, let's have a look at Crikey.com.

Finally, while there appears to be little hard evidence (and the calculations would presumably be complex at best), some have suggested that there may be some environmental benefits from mobile phones and associated items such as tablets (Berry and Goodman 2006; Ling and Bashir 2011). These potential benefits include energy savings through: using mobiles to remotely monitor and manage energy-using appliances; creating more efficient transportation systems, for example by facilitating better coordination of people and transport and by providing real-time information on traffic and demand for services; reducing travel needs through electronic commerce, video-conferencing and other forms of mobile transactions (Chapter 11); substitution for other electronic items as phones increasingly take on multiple functions; and substitution of phones and tablets for books, DVDs, CDs, magazines, and newspapers (Chapter 13).

WHAT TO DO?

Most of the literature on environmental impacts of mobile phones highlights industry strategies as the key pathways to improved sustainability, particularly as network infrastructure grows and consumes large amounts of energy and materials. Such strategies include design innovations to reduce energy and material use during manufacturing and to facilitate recycling; modular phone design to facilitate component upgrades; the use of software upgrades to maintain/enhance phone usefulness; sharing of network facilities between companies; provision of more efficient and universal electrical chargers; and institutions and incentives to improve phone recycling through pathways that are easily accessible, and which effectively and safely recycle materials. The evidence discussed above also highlights the role of phone companies in the design of contracts, network access, and in the provision of new phones as well as the need to address issues of data privacy at the time of phone replacement. Many of these issues are not ones that can be addressed within the limits of everyday life, except through non-participation in phone ownership. Rather, collective action is required by industry and government.

It is also clear that, for an individual who wishes to function within a society in which mobile phone use informs and structures family, social, and business interactions, not owning a mobile phone will come at a cost. Mobile phones and their various applications are central to myriad daily interactions by text message, social media, and phone calls themselves. 'Micro-coordination' (Ling and Bashir 2011) of many aspects of daily life for families, friends, and businesses occurs via mobile phones. In addition, they are, of course, becoming increasingly important to the consumption and distribution of all kinds of media and news, and as discussed above, they are important to individual identities through, for example, the fashion status and iconography of one's phone and its contents.

For mobile phone owners who wish to sidestep this dilemma, what are the options? The first possibility is to 'resist' fashion, the lure of the new and of enhanced technology, and the phone companies' incentives, and not replace one's phone until its functional life, or at least its practical useful life to the owner, has been reached. Hanging onto a phone for more of its functional life could add up to eight years to its use according to the evidence presented above, reducing dramatically the demand for raw resources and energy used in manufacturing and transport of new phones. For this to occur, we suspect that the sensibilities of that rare maverick, the excessively proud owner of an old phone, would need to become far more common. Take Margaret, from our ethnography:

> Oh only [replace my phone] when it's really dead, when I've dropped it, smashed it, all the buttons have fallen off. Because I'm not terribly attached to them I don't feel the need to update it regularly. It's only when it's hard to work out what buttons are anymore or it's held together with an elastic band that I ask if I can get it changed. It's usually somebody else that looks at my phone with utter disgust before me and says, I think you should really get another phone. I say, I can still use it.

Incentives to replace also need to be curtailed greatly. It may also require phones that are more easily upgradable without requiring wholesale replacement. Both these options work against the interests not only of the phone manufacturers, but also potentially of used phone refurbishers, whose profitability often requires relatively new models for resale (see Geyer and Blass 2010, who also suggest that phone reuse by first-time users can be in the phone manufacturers' interests insofar as it develops future phone markets). The second option for phone buyers is to look for those phones that have been manufactured with environmental and end of life practices squarely in mind. For phones, as in many consumer arenas, there are purchasing guides such as that on the GoodGuide website (GoodGuide 2012), which rates phones on environmental criteria. Phone company websites may also actively promote the 'eco' features of their products or of particular models, though as with clothes (Chapter 3) and cars (Chapter 10) the risk of greenwash is ever-present. The third thing to do (at the time of replacement) is to recycle old phones through recycling programmes that safely and legally recycle phone materials for reuse, and do not transport them to countries without adequate e-waste recycling and disposal systems. In many countries there are such schemes run by phone companies or by telecommunications industry associations. The fact that recycling rates nonetheless generally remain low despite the existence of a profitable reuse/recycling industry (Silveira and Chang 2010) suggests that there is much further work to be done.

15. Solar hot water

In more affluent countries hot water, like toilets (Chapter 6), mattresses (Chapter 8) and refrigerators (Chapter 12), is one of those things that gets noticed more when it fails than when it works. When, as is often the case for hot water systems, they reach the end of their useful lives suddenly, property owners find themselves having to make a quick decision about a replacement system. In the past the default in many countries was a standalone electric hot water storage system, composed of an insulated tank heated by an electric element. Installing such high energy use, high greenhouse gas producing systems took little thought, except that needed to find the plumber's phone number. This is no longer the case. Governments of various persuasions have targeted hot water heating as a major area of reform for reducing greenhouse gas emissions. In Australia, from 2010 the federal, state and territory governments (except Tasmania) are working together to phase out greenhouse-intensive (electric) water heating systems, with the aim of decreasing emissions by 51.1 million tons over ten years from 2010 to 2020. This is the equivalent of taking 1.4 million cars off the road during the same period and should deliver 4 per cent of Australia's current projected greenhouse gas abatement by 2020 (DCCEE 2012a). Property owners are now forced to consider a range of alternative hot water systems. In Australia as elsewhere, the options include several different solar hot water system designs, gas instantaneous systems, air source heat pumps, and gas storage systems. The choice is yours.

Of the options, solar systems offer the highest potential energy savings and such savings are enthusiastically promoted by governments and sellers of solar systems. Installation, particularly in China, is growing (Weiss and Mauthner 2012). Yet despite the long availability of solar hot water technology, uptake has not always been as significant as the potential benefits might suggest. Even in Australia, one of the leading adopters of solar hot water systems (Weiss and Mauthner 2012), only 8.5 per cent of households had solar water heating in 2011 (ABS 2011b), a slow rise up from 5 per cent in the mid-1980s (Foster 1993). The United States has a high proportion of total global installed capacity in solar hot water systems but lags other countries on a per capita basis; it does not feature among the top ten countries for new installations per capita (Weiss and Mauthner 2012). In New Zealand only about 3 per cent of households have solar hot water systems (Grieve et al. 2012). Installation rates

vary according to factors such as the history of fuel availability and the relative costs of energy sources, as well as regulation. In Israel, where acute fuel
shortages occurred in the 1960s and 1970s and where solar hot water systems
are required by the Israeli government, 90 per cent of households use such
systems (Perlin 2002). In Australia, installation rates vary from 54 per cent in
the Northern Territory to 5 per cent and 3 per cent in Victoria and New South
Wales respectively (Clean Energy Council 2011). Perlin (2002) attributed the
variation to the availability of relatively cheap coal-fired electricity production
in east coast states.

SUSTAINABILITY ISSUES

Alternatives to electric storage systems can save energy and greenhouse gas
emissions. In Australia, where about half of households still have electric storage heaters (ABS 2011b), heating water is considered the 'largest single source
of greenhouse gas emissions...and accounts for about a quarter of household
energy use' (DCCEE 2012a). In the UK the predominance of electric heating
systems means that as much as '56% of the average UK primary CO_2 footprint
can be accounted for by water heating' (Hackett and Gray 2009, 37–8).
Similarly, in New Zealand, where it is estimated that 80 per cent of households
still use electric storage systems, hot water heating accounts for approximately
30 per cent of total household electricity consumption (Grieve et al. 2012). Hot
water can be saved by changing water use in the home, such as by reining in
excessive showering habits, the measurement of which has moved one group
of researchers to label showering a 'leisure activity' (Makki et al. 2012) (see
also Chapter 7). More efficient water heating systems, however, can also reduce
energy consumption for water heating by a significant amount. In this chapter
we will focus on issues around the choice and installation of solar systems
themselves, while recognizing, as we do elsewhere in the book, that much hot
water use is governed by factors such as ideas about cleanliness, the role of
bathing or showering in everyday life, and household composition (Wilhite et
al. 1996; Shove 2003; Troy et al. 2005; Patton 2011).

There are two basic types of water heating systems – storage and instantaneous (Riedy and Milne 2010). In storage systems, water is heated and stored
for use when required. Storage heaters are effective in any climate, however
some heat is lost through the walls of the tank (DCCEE 2012a). With storage
systems, the stored hot water can run out, and due to concerns that this may
happen, tanks are often oversized and overheated, increasing energy usage
(Riedy and Milne 2010). Instantaneous (also known as 'continuous flow')
heaters only heat the water when it is required and do not use a storage tank.
These systems can save energy and household costs, do not have heat losses

from a tank and the absence of a storage tank saves space around the house. Each of these two types can use a variety of energy sources, including two in combination. The options are in order of increasing energy efficiency: electric storage (4.2 CO_2-e), solar with electric booster (1.4 CO_2-e), gas instantaneous (1.2 CO_2-e), air source heat pump (1.2 CO_2-e), gas storage (1.1 CO_2-e), and solar with instantaneous gas booster (0.4 CO_2-e) (DCCEE 2012a). Air source heat pumps are essentially air conditioners in reverse and work by extracting heat from the surrounding environment (which can be air, ground or water depending on purpose and setting) and transferring that to water in a tank. Solar with instantaneous gas booster produces under half the greenhouse gas emissions of the next most efficient system, gas storage. Even the least efficient alternative to electric storage, solar with electric booster, produces one third of the greenhouse gas emissions of electric storage. One of the attractions for a household of solar hot water systems is thus the potential to save on household energy costs. Payback time varies according to climate, hot water use patterns, and energy costs but, in Australia, they are estimated to be from five to ten years (Riedy and Milne 2010). A recent estimate for Pakistan was six to seven years (Asif and Muneer 2006). In cooler climates, the payback can be longer. In the UK the average solar hot water system costs £4800 (Energy Saving Trust 2011a) and annual savings are typically £55 if replacing a gas system and £80 if replacing an electric system (Energy Saving Trust 2011b). Even taking into account the cost of alternatives, the economic payback time is likely to be considerably longer in the UK than in places with higher levels of solar irradiation.

Solar hot water systems have a history going back to the late nineteenth century. The first commercial system was in use in the United States by the 1890s and design features that continue to be integral to modern systems were evident in a commercial design available from 1909 (Perlin 2002). This technology was so successful that by 1941 half of Florida's population used solar hot water systems until they were out-competed and out-marketed by suppliers of electric hot water systems, including electricity utilities (Perlin 2002). Today a variety of domestic solar hot water heaters is available and the applications of the technology are also expanding into space heating (Weiss and Mauthner 2012) (Chapter 5). While heat pumps are often classified as solar systems and are highly efficient, the focus here is on the best-known solar hot water systems, those with the familiar roof top 'collector' panels and a tank to store hot water. Such systems come in a variety of configurations but they all essentially consist of dark-coloured collectors that absorb heat and through which fluid (water or, in cold climates, antifreeze) passes to be heated. Heated water is then stored in a tank to be accessed by the household. Solar systems cannot always supply all household hot water demands. They are therefore fitted with gas or electric boosters which heat water. One of the most common designs is the roof-mounted flat-plate thermo siphon system, in which dark-coloured absorber plates in an insulated box are attached to a tank. Liquid in the panels heats up and circulates in tubes, transferring heat to water in the tank. Evacuated tubes are an alternative type of collector and are more efficient but more expensive than flat plate collectors.

TRANSFORMATIONS AND COMPLEXITIES

The information above suggests that choosing a solar hot water system should be relatively straightforward, particularly if one is motivated by a desire to reduce household environmental footprint as much as save money. This is a common factor in the installation decision, as we found in a series of interviews we conducted with solar hot water system owners in our ethnographic collaboration with Peter Osman (CSIRO). For example, one said:

> the main reason was – I suppose, issues with climate change...doing something to make a difference...less so, because I knew it was going to take a long time to pay back, to reduce the bills. But more just to be doing something environmentally sound.

Nevertheless, low rates of installation in some jurisdictions suggest that the decision may not be so simple. This was the finding of a New Zealand study based on interviews with householders in which its authors found a range of barriers to installing a solar hot water system (Grieve et al. 2012). Barriers included up-front costs and uncertainty regarding which choice to make despite an apparent abundance of information. In Australia, such factors and a tendency to only replace hot water systems at the point of failure means that people tend to purchase the same kind of system as they already have (George Wilkenfeld and Associates 2010).

The high capital cost of solar hot water systems especially prevents their installation. Solar systems, including heat pumps, have the highest up-front costs (although governments have varying rebate schemes). In Australia, the capital cost of gas boosted solar is over twice that of the next most efficient system, gas storage, whereas the running costs of gas boosted solar are under half those of the best of the other options (Clean Energy Council 2011). Lifetime costs are generally comparable. Given that solar systems are expected to have service lives of 12 to 14 years (George Wilkenfeld and Associates 2010), the lifetime cost benefits to a household may not be as large as commonly perceived (Ministerial Council on Energy 2008).

Cost is an issue even if solar hot water is preferred for environmental reasons. High up-front costs amplify uncertainties and risks associated with a poor choice (Grieve et al. 2012). Grieve et al. (2012, 4) conclude that not only must the householder be cognizant of the solar option and its potential, and be able to afford it, but that they must deal with the 'critical barrier' of making sense of information and applying it to their specific circumstances.

Online forums and our interviews revealed positive stories about solar hot water, but also reflected uncertainty and opacity of information. A contributor to an online discussion of solar hot water facilitated by Australian consumer organization, Choice, wrote:

> I just wanted to replace my old ... continuous power electric storage heater which used about 2 kWh per day but when I enquired I learned this is not allowed and now I am immersed (forgive pun) in this eco nightmare where I have to a) find an extra 2 or 3 grand and b) become an overnight technological expert. I am prepared in principle to do my bit for the environment but it seems this technology is expensive, still unreliable and not altogether satisfactory. (Barnes 2008)

Such forums also contain accounts of less-than-optimal experiences with solar hot water, ranging from disappointing performance to rusting gutters due to warm overflow water being continuously directed into them. Our interviewees also experienced difficulties making sense of information about which hot water system in general and which solar system configuration in particular. One relatively content solar system owner experienced problems making sense of information from various sources and described a sense of unease that he may be getting an inferior product that he had no way of validating:

> There was information but ... you just never know how true it is ... these people that I rang, about the evacuated tubes, they said "Oh yeah our product is excellent, more efficient". But ... I didn't feel hundred per cent confident we were getting the whole picture ... I also went to Future World [an alternative technology centre] 'cause I'm a member of Future World and I listened to a guy talk. He was talking about how in China, you can buy these really cheap from the side of the road.

What does this mean for households? Information that aims to encourage homeowners to install solar systems is plentiful, but households do the hard work of figuring it all out. Governments the world over have established websites for this very purpose, such as the Australian government's 'Living Greener' website (DCCEE 2012b) which aims to provide practical information on living sustainably. Consumers are encouraged to find out which system best suits their household size, location, house configuration, and climate, to work out the relative installation and running costs of different systems, to conduct research on these issues and to seek 'expert advice'. This is clearly a lot to ask of people for whom hot water heating is background infrastructure. It is also difficult for potential purchasers to get beyond advice that one's choice will 'depend' on their specific situation or that the main source for 'expert' advice is installers of solar hot water systems. At this point, the process can halt and the default choice can be too swiftly made (Grieve et al. 2012). Homeowners who persist and seek information and quotes from installers subsequently run into issues of trust. Grieve et al. (2012) found that homeowners were concerned about receiving biased advice from installers, a theme echoed in our interviews. One homeowner who had conducted signifi-cant research himself and who ultimately relied more on the experiences of other homeowners to make his choice said:

> the first guy didn't know what he was talking about, really. I knew probably more, and I thought, you're really trained in selling this based on the rebate, and you are a salesperson, and you don't know the ... [science]. I will ask the difficult questions of anything that I buy or install.

Beyond issues of choice and information, how well do solar hot water systems meet their intended purposes (providing hot water, saving energy and money, reducing greenhouse gas emissions)? The Australian federal government advises that solar hot water systems can provide between 50 per cent and 90 per cent of water heating needs (DCCEE 2012a). A UK study that involved detailed monitoring of 88 domestic solar hot water installations found that well-installed and properly used systems could provide about 60 per cent of a household's hot water (the highest was 98 per cent) and that the median was 39 per cent (Energy Saving Trust 2011b). However, the same study found that if the system was poorly installed or used, the contribution from solar was as low as 9 per cent and that household hot water energy costs could even rise (Energy Saving Trust 2011b). This is consistent with preliminary analysis from our study of household solar hot water use. Households with electric boosted systems had highly variable (by as much as a factor of 20) off-peak electricity consumption per person, a variation not easily explained by other aspects of the household. In the UK study and in our research, key variables include installation (configuration of the booster and when it is set to turn on), inadequate system capacity relative to household size, high temperature settings, lack of advice to households regarding use strategies, and a lack of pipe insulation. Household factors include poor timing of hot water use such that the booster is used excessively, and uncertainty as to how to run the system to maximum effect. One simple issue was a lack of understanding of how the system operated. This led to uncertainty in controlling the booster, including where the switch was installed:

> Experimenting and then finding that we don't have hot water in the morning is a pain in the butt. So you try not to, it's not an experiment that has a neutral result to it. So we did it just for a very, very short time, because we didn't know. We were all keen about it, because now that we've got this switch here and we switch it on and

off, and we are going to be able to save a lot of electricity here. That's gone out a little bit, probably because... we've lost interest in it. We didn't know what it was actually doing, we didn't want to not have the hot water.

Other households had similar trials. One had conducted such experimentation and had ended up with frequent cool water even in summer when the solar system should have provided virtually all hot water needs. The household had then reverted to the same default strategy, rather than pursuing why the system was not functioning optimally.

WHAT TO DO?

The decision to install a solar hot water system is not as straightforward as one might expect. They do have the potential to save on energy use and greenhouse gas emissions and thereby provide a public good. Financial benefits for individual households are less clear, ranging from reasonable to marginal depending on location, use, and installation. Installation is more likely to be a non-financial decision, motivated by environmental concerns or, in some locations, regulation. Given the cost, uncertainties and hesitation are barriers to installation. Moreover, there may be simpler and more certain ways to reduce household greenhouse gas emissions. The UK Energy Savings Trust found that the typical greenhouse gas emissions savings from installing solar hot water were about the same as could be achieved by 'draught-proofing round all the doors, windows and skirting boards in a gas heated or electrically heated home' (Energy Saving Trust 2011b, 19). Such a strategy in conjunction with switching to gas storage or instantaneous system or a heat pump would generate significant greenhouse gas emissions savings.

Beyond such strategies, there is a clear need to provide far better, location- and household-specific information that is accessible without recourse to an installer. Most information at the moment points out the issues to consider; there is little guidance on how to take the next step. Wright et al. (2009) contains more practical guidance.

Consistent with our argument that household sustainability is as much about the relationships outside the household as about the action and nature of the household, the role of installers as gatekeepers also appears to be critical. Not only is there the problem of trust. Both our interviewees and the respondents to the UK study received variable information and assistance with operating their hot water systems. Some of our interviewees were willing to engage with the systems and experiment, willing to change the status of the hot water in the household and to make it a more visible and active component of their everyday lives, but they were quickly discouraged. This was at least partly because they lacked information about how the system worked and they were limited in their ability to then apply such knowledge to their household circumstances. Installations also vary in quality. Pipe insulation is a key issue, now required in the UK, but not necessarily elsewhere. The UK Energy Saving Trust report drawn on in this chapter provides a checklist of what households should expect from an installer.

Finally, it is worth mentioning the likely development of 'smart' controllers that have potential to manage booster operation in more nuanced ways. Such controllers will potentially adjust booster use according to variables such as weather and water use patterns, not just on the basis of temperature. While this may perpetuate the status of hot water systems as background, unnoticed infra-structure, it may also be a mature response to the limits to household interest in engaging with their hot water heater.

16. The garden

Domestic gardens occupy contradictory positions in sustainability debates. On one hand, suburbs themselves are cast as environmental nightmares, using the pejorative 'sprawl' and the associations this evokes: heavy car dependence, high infrastructure costs, larger dwellings than necessary. On the other, gardens provide a variety of opportunities for environmentally beneficial activities: growing food (Chapter 2), capturing and storing water (Chapter 4), drying laundry (Chapter 7), enhancing biodiversity, waste disposal, compost. Suburban experiments such as Happy Earth share ideas about changing 'a typical suburban house… into a sustainable, healthy home and organic food garden' (www.happyearth.com.au). Conceptually, the garden dilemma is part of the wider debate over how much humans should be 'contained' in urban fortresses in order to protect 'nature' somewhere else, 'out there'.

Gardens, particularly lawns, cover about one quarter of the urban land surface in the United States, and are a major force for ecological change (Robbins and Sharp 2003; Morris and Bagby 2008). The garden care industry is huge. Americans spent US$8.9 billion on lawn-care inputs in 1999, and this has been growing steadily (Robbins and Sharp 2003). Robbins and Sharp showed that the total quantity of lawn chemicals being applied to American lawns (9.127 kg/ha) is far greater per hectare than in agricultural fields (3.238 kg/ha). Morris and Bagby (2008, 226) compared the pollution pathways and environmental costs of two different, but typical, patterns of care: 1) a 'conventional' system of petroleum-based fertilizers and pesticides, a gasoline-powered lawn mower, and substantial irrigation to maintain a green, weed-free, lawn and garden, and 2) an 'organic' system using backyard compost to provide nutrients, supplemented moderately by purchased non-synthetic soil amendments, an electricity-powered mower, no pesticides, and drought-tolerant species. With benefits estimated in dollar terms, they concluded that the environmental cost of the conventional system was nearly ten times that of the organic.

Another increasingly common garden feature, the swimming pool, also has significant implications for the use of water, chemicals and energy. Forrest and Williams (2010) found that the environmental impact of pools varied widely between nine US cities, influenced mostly by climate. In arid-zone Phoenix, pools accounted for 22 per cent of a household's electricity use and 13 per cent of its water use. They also found that community pools can yield substantial reductions in both. Backyard pool occurrence was much lower in subdivisions with a community pool, leading to 60 per cent less swimming pool water and energy consumption per household than in subdivisions without a community pool (Forrest and Williams 2010, 5601, 5605).

The somewhat polarized debate around urban form (e.g. Troy 1996, Newman and Kenworthy 1999, Williams et al. 2000, Frumkin et al. 2004, EPA 2006, Gleeson 2008) has led to a wider acceptance of urban intensification as a more sustainable solution than endlessly increasing suburbs, but the situation is more complex than this. Urban consolidation risks biodiversity loss and a variety of effects associated with increased surface sealing (Pauleit et al. 2005; Hasse and Nuissl 2007). In many affluent countries, and particularly in Australia, the separate house and garden persists as a highly valued form by its human inhabitants (Gleeson 2006; Head and Muir 2007a; Davison 2011). Residents report that outdoor spaces enhance social networks, connect to and replicate distant cultures, allow them to positively interact with nature, contribute to aesthetics and improve quality of life (Head and Muir 2007a).

Whatever the range of views about domestic gardens, they are embedded in our sustainability decisions over both short and long timescales. In Australia, for example, only 1–2 per cent of building stock is replaced or added to by new-build each year, so it is important that we think in terms of retrofitting what we have, as well as imagining new forms. Over the longer term these annual changes compound considerably. So, for the 30 years between 1990 and 2020, the number of occupied residential households in Australia is forecast to increase from six million to almost ten million, an increase of 61 per cent (DEWHA 2008, ix).

In the timeframe that we need to mitigate or adapt to climate change, gardens will be part of the picture, and research in a number of parts of the world has responded to this. Gardens are a dominant land-use type in urban areas, occupying up to one-third of land areas (Gaston et al. 2005, 3327; Mathieu et al. 2007, 179). Urban tree canopy cover can protect environmental quality and human health through carbon sequestration and storage and air quality improvement (Nowak and Crane 2002, 382). The 'Biodiversity in Urban Gardens in Sheffield' (BUGS) project in the UK has shown that larger gardens can support a variety of land-cover types that are important for achieving improved ecological and environmental functions (Smith et al. 2005; Loram et al. 2008). A UK study that considered the environmental parameters of surface temperature, rainfall runoff and green space diversity demonstrated close links between green space provision and environmental performance potential (Pauleit et al. 2005, 307). Gardens will necessarily be part of the mix.

BIG GARDENS, SMALL GARDENS OR NONE?

Of course, suburban gardens are themselves variable. Ghosh and Head (2009) compared 'traditional' (relatively smaller houses and larger blocks of land, widely built in the decades post-World War II) and 'modern' (larger houses and smaller gardens, becoming more common in the last 20 years) suburbia against several sustainability indicators. They found that the traditional suburban configuration had higher potential for connectivity of green space, and for food production. Existing tree canopy covered 23.5 per cent of traditional blocks, compared with 5.6 per cent of modern ones. Modern developments had comparatively younger fast-growing trees which could enhance carbon capture potential in the near future, while traditional blocks had stocks of mature trees. The food production potential of traditional blocks was significantly higher than modern blocks due to larger land areas.

In contrast, modern blocks had higher potential for water collection through more extensive roof space; the modern block could supply 135 per cent of its total (potable and non-potable) water requirements, compared with 63.4 per cent for traditional blocks. Realizing this water potential usually requires the necessary tank infrastructure to be provided for in the building stage, as it may for example need to be underneath the house (Chapter 4).

Regardless of garden size, over 90 per cent of Australian households in detached dwellings have an outside clothes line (Head and Muir 2007a, 32), an important means of minimizing electricity consumption. Of course, people living in apartments have much less opportunity to dry their clothes outside; indeed it is often prohibited to sully balconies with flapping underwear and shirts, and in the United States some low density suburban housing estates ban clothes lines on aesthetic grounds (Chapter 3).

POTENTIAL AND PRACTICE

The form of the garden and house are not the only variables. Gardens are not just blank slates or pieces of land to which things are done, and we should not only assess them as resources for land, soil, water and growth. Human behaviours and practices are important parts of the equation. In our survey research, we found mainstream sustainability practices (such as reducing indoor water consumption, moderating purchasing decisions, recycling and switching off lights) to be statistically more frequent among households living in detached houses than those living in units (Waitt et al. 2012b). As we saw in Chapter 4, the house/garden configuration provides many opportunities for the householder to be conscious of, and intervene in, the process of provisioning, storage, transfer and waste disposal of water. In contrast, apartment dwellers usually have less control over the shared infrastructural systems. As noted by Gleeson (2008), higher-density unit dwelling often entails higher expenditure on goods and services (e.g. swimming pools, lifts and common property lighting) while strata laws and the need for consensus among a large number of cohabiting households limit autonomy to reshape the systems, practices and habits surrounding waste management and utility use in apartment blocks. The flipside of this is that sharing some forms of appliance consumption, such as washing machines (Chapter 7), and discretionary items, such as DVDs and books (Chapter 13), can be enhanced in higher-density living.

Complicating matters further is that the *potential* of gardens, as discussed above mainly in terms of land area, can be rather different to the actual outcomes, which depend on a variety of factors. Not least of these is the amount of work involved, and its connection to wider patterns of domestic and labour force activity. Young mothers in our ethnographic study described themselves as 'always at the clothesline', indicating that whatever else the garden is, it is also a workplace. To take another example, vegetable gardens require both start-up investment of time, and ongoing inputs. Although more than half of households grow some vegetables and/or herbs in their garden (for example 52 per cent of the 265 backyards studied by Head and Muir 2007a, 90), only a few are self-sufficient in these.

THE SOCIALITY OF THE GARDEN

Patterns of sociality in and around gardens are themselves diverse, and they also have diverse environmental implications. Positive patterns of conviviality and human sharing are documented in our research. Gardens are places from which abundance – of produce or flowers – is shared across the fence, often maintaining long-standing generational patterns. As Sylvia commented:

> The other great pleasure is that I'm growing things both for myself and for my neighbours and friends and it's wonderful to be able to share things with people who either haven't got a garden or haven't the time to cultivate things... My father would lean over the fence and say to the elderly couple next door, "have a punnet of blackberries" or "have you got any gooseberries, would you like some of ours, have some apples". (quoted in Head and Muir 2007a, 93)

For many households, the garden is the place where most entertaining occurs, as meals are shared at the barbecue, in informal outdoor living and dining spaces.

But things are not all lovey-dovey in the garden. It is also a space where urban sustainability conflicts are writ small. Trees and lawn provide two examples. No matter how attractive the idea of creating urban habitat corridors that join up the vegetation of different blocks (Parker et al. 2008), we know that any attempt to foster more widespread tree planting in suburban gardens would face significant cultural barriers. Garden trees are a major source of conflict and generate considerable antagonism between neighbours, who talk about love and hate, mess and danger. In Head and Muir's (2007a) ethnographic research some even confessed to having killed their neighbour's trees. Others talk of loving trees which may be invasive weeds.

Robbins and Sharp (2003) argued that one of the contributing factors to the dominance of lawn in the United States is a particular 'moral economy' in which the signification of competitive neighbourhood status, the nuclear family and a particular view of environmental health are all associated (along with the economic imperatives of the lawn-chemical industry itself) (see also Robbins 2007). Lawn is a good example of a garden form that has different cultural meanings in different societies. In our Australian research, the lawn usually represents order and tidiness to some degree, but the moral association with respectability is not as strong as documented for the United States (Robbins 2007). Or at least it is more contested, with a polarity of evidence between lawn lovers and haters (Head and Muir 2007a). While lovers talk of 'the sensory pleasures of grass, the pleasure of the labour of mowing and lawn's importance as a play area for children', haters see lawn as an ecological evil, 'a voracious consumer of water, chemicals and time better spent on other things' (Head and Muir 2007a, 64). For example, contrast Jim describing the turf in his relatively new garden:

> Everyone that comes here for a barbecue usually takes their socks and shoes off and stands on it and they say it's great.

with lawn hater Heather:

> I can't understand why people water their lawns but they do and it serves no useful purpose unless you've got something grazing on it. (quoted in Head and Muir 2007a, 65, 66)

Contrary to expectations, a surprising number of men described enjoying the therapeutic nature of mowing the lawn, the smell of the cut grass, and the cold beer at the end! Others, sometimes but not always female, described themselves as being on a 'mission' against lawn.

GARDENS AND CHILDREN

Lawn haters often qualified their antipathy to lawn with the statement that they do not have young children at the moment. This reflects a commonly expressed feeling that kids need lawn. Rebecca, in the middle of building a house, commented, 'All I want is for [my son] to play on the grass because he's had no grass for a year' (Head and Muir 2007a, 66). The view that children need gardens is very widespread among adults, expressed in having somewhere safe to play, and the moral imperative of 'getting outside' rather than being connected to electronic screens all the time (cf. Chapter 13). Head and Muir (2007a, 44) referred to this as 'the moral power' of the backyard. For example young mother Julie described the garden space as an important criterion in her house purchase:

> we were very aware when we bought the place that we wanted the kids to use the backyard. And we wanted it to be a place where hopefully they would bring their friends where we could have a lot of room and be able to keep an eye on them so that they weren't wandering around the streets.

A different, and much less common view was expressed by Alan, whose fascinating Alice Springs garden (in an arid environment) was filled with adventurous play options and lots of junk, in a way that would contravene most occupational health and safety guidelines:

> I believe in the dignity of risk for kids, and risk taking and physicality and learning your own limitations. It's a big part of my personal philosophy on rearing kids I guess. (quoted in Head and Muir 2007a, 44)

In the same study, very few parents talked specifically to us about children and nature. Children growing vegetables, playing with or tending domestic animals, and observing other creatures were mentioned much less frequently than the importance of open space and safe physical activity. This finding is interesting given the importance that research attaches to connecting with the natural world in the first few years of life (Measham 2006; Cheng and Monroe 2012). Further, this is not to suggest that the residential garden is the only place for urban children to connect to the non-human world. Parks and public spaces are important in high-density urban contexts such as Singapore (Kong 2000), and many parental anxieties that play out there parallel those in the suburban contexts of Australia and the United States.

HUMAN AND NON-HUMAN VALUES OF GARDENS

Ethnography makes clear that gardens are complex co-productions between humans and non-humans – plants, animals, birds, rocks, soil and rain. They are one important place where human (sub)urban dwellers can enhance their awareness of the wider webs of life. Cultural research into gardens also alerts us to the dangers of focusing on physical form while ignoring the social and ecological relations that constitute and interact with that form (Power 2005; Head and Muir 2007a). For example, Lynette exemplified widely shared feelings in our backyard study, with her description of her backyard:

> I just get a feeling of tranquillity and serenity and it's just a little haven, especially when I go out sometimes and watch the frogs … you can just watch them and it gives me a great feeling of satisfaction because it's something that sort of puts you back in tune with nature which is sadly missing from our society and most of our lives. (quoted in Head and Muir 2007a, 71)

Gardens help demolish the myth that urban dwellers are uniformly alienated from and antipathetic to the non-human world:

> Outdoor domestic spaces are highly valued environments for the privacy and freedom they offer, and for the interactive relationships they facilitate with family, friends, pets, birds, other animals, sunsets and breezes. Connections are developed through habit and close observation at different times of the day, in different seasons, over different years. (Head and Muir 2007a, 155)

Bird visits were the most talked-about animal activity. Although relatively transient visitors, birds afforded great pleasure to the keen and casual watcher alike. For example, Jessie of Wollongong said:

> I talk to lots of people who are like me, go out into your garden and look around and watch the birds and see what happens, you forget if you've got any stress. I think everybody needs a backyard. (quoted in Head and Muir 2007a, 142)

Gardens do not have to be big to be effective in the role of facilitating connections with non-human nature. A number of participants in our study had only balconies or small courtyards, but could still talk with passion about changing cloud patterns or spectacular flowering pot plants. The intensity of engagements that people develop with a whole garden can be focused on one or two plants. An important pointer for the future is that a 'desire for outdoor connections can be met in diverse ways. Size matters, but not in the same way to everyone' (Head and Muir 2007a, 44).

WHAT TO DO?

Gardens are neither an environmental villain nor an Arcadian panacea in themselves. They are places of variable human behaviours and complex interactions with the non-human world. Retrofitting of existing gardens and thinking creatively about new forms will both be important in the decades ahead. Some things are no-brainers, such as encouraging the outdoor drying of clothes in places where the climate permits (Chapters 3, 7).

People place high value on having some private and autonomous outdoor space, but it does not necessarily need to be huge, provided there is access also to shared public green space. Nevertheless large gardens do provide opportunities not available in small spaces. Variable housing and garden styles are needed in different cultural frameworks, and at different stages of the life-cycle. We can do much more to make these configurations flexible and adaptive.

There are many ways in which connections between the gardens of individual households could enhance sustainability at the neighbourhood scale, for example in food production, water collection and storage, and tree planting. In considering these possibilities there needs to be at least as much attention to the social mechanisms of connection as to appropriate design planting solutions. As much effort needs to go into nurturing communication throughout the suburbs as in nurturing the trees.

17. Christmas

Christmas has been described as the world's greatest annual environmental disaster (Bryant 2009). Christmas decorations, gift-wrap, cards, food and gifts contribute significantly to intensive resource consumption and waste, particularly among high-income groups in the West.

Understanding sustainability dilemmas associated with the contemporary Western Christmas requires an appreciation of the festive season as much more than a religious ceremony; it is also a cultural event of immense economic significance. Christmas has a long history and contemporary cultural value as a celebration of material abundance, with roots stretching back at least to pre-Christian Roman festivals (Miller 1993). Christmas is a treasured period away from work, a welcome annual foil to daily routine. It is a time to affirm family with feasting, and practise generosity with gift-giving. It is also a time for upholding wider social ties, through sending cards, distributing and receiving gifts, and sharing meals. Focus turns to giving to children and those who are less well off. Given Christmas' focus on material abundance, attempts to reduce Christmas gifts, decorations and food in the name of sustainability confront the cherished 'Christmas spirit'.

Yet Christmas is also a source of significant stress. Householders worry about the affordability of Christmas, over-spending, the influx of excessive stuff arriving in homes in the form of gifts, maintaining the Christmas spirit in the face of commercialization, eating too much, and family conflict. It is no coincidence that the modern form of Christmas emerged at the same time as industrial capitalism, the time when material items became mass-produced and consumed for the first time (Carrier 1993). In the current economic system, capitalism is increasingly dependent on the Christmas season (Basker 2005). While attracting some anti-consumer protests, Christmas is the most intensive trading period of the year. As Mansvelt (2008) argued, households receive mixed messages about their role in the economics of Christmas: they are enjoined to spend more to support retailers, and hence the national economy, but not to finance their expenditure through excessive debt. Further, excessive material things, such as elaborate Christmas lighting displays, wasted food and unwanted gifts become a lamented feature of Christmas rituals: materialism has become a global Christmas cliché (Carrier 1993; Miller

1993). Yet, as we show in this chapter, some degree of material abundance remains fundamental to the celebration of this and other rituals.

SUSTAINABILITY ISSUES

The sustainability issues associated with Christmas are far reaching, involving domestic practices as well as global agricultural, transport, manufacturing and waste industries, which produce, distribute and dispose of the range of goods commonly purchased for Christmas meals, gifts and decorations, such as turkeys, alcohol, clothes, books, films, music, cosmetics, fragrances and Christmas lights (Haq et al. 2007). Christmas is a period of intensified consumption, not simply mirroring but compounding many everyday sustainability issues, such as those associated with food, travel and energy use considered elsewhere in this book. Yet surprisingly little research has been done to quantify the specific environmental impacts of Christmas. One British study calculated that Christmas accounted for 5.5 per cent of annual household carbon dioxide emissions, from a time period less than 1 per cent of the year (Haq et al. 2007). In the UK, £4 billion was spent on unwanted gifts each year. If unwanted gifts were not bought in the first place, the Christmas shopping footprint would be reduced by 80 kg CO_2-e per person (Haq et al. 2007). Common household practices also play a part: putting vegetable peelings from the Christmas meal in the regular bin instead of compost, for example, contributes an extra 7kg CO_2-e per person to a typical Christmas meal. However, these figures focus only on carbon dioxide emissions. They do not take into account other impacts of Christmas, for example, on water usage or waste tonnage. Even the climate impacts may be underestimated: a study of the greenhouse emissions of a typical Christmas meal noted that carbon dioxide accounted for only 42 per cent of the impacts of a Christmas meal, with the remainder contributed by nitrous oxide and methane (Stichnothe et al. 2008).

CHRISTMAS GIFTS: IT'S NOT JUST THE THOUGHT THAT COUNTS

Notwithstanding potential economic savings for households, there are deeply embedded cultural values that inhibit shifts to more sustainable Christmas rituals. Pleas among family for 'no more stuff!' and 'no gluttony this year!' clash with the treasured art of giving, and the joy of generous, festive meals shared with extended family and friends. From an anthropological perspective,

Christmas can be understood as a significant annual ritual for negotiating one's place in the world (Levi-Strauss 1993). Children in particular experience a rite of passage into adulthood when they discover the 'truth' about the non-existence of Santa Claus. Before this, children are expected to practise and express gratitude and generosity, in domestic good behaviour throughout the year, while having their role as 'innocents' affirmed. They are then taught that they are the deserving recipients of a wider generosity, embodied in the well-known figure of Santa Claus. When the truth about Santa is revealed, children are expected to relinquish their identity as innocents and perpetuate the myth for the next generation of children. However, the exchange of gifts between adults suggests that childhood belief in a generous world is never fully relinquished, despite the power of the commercial economy in which contemporary Christmas rituals are so deeply embedded (Levi-Strauss 1993).

Indeed, attachment to a belief in a wider generosity is strongly associated with the practice of exchanging Christmas gifts, which in contemporary times, is inseparable from Christmas shopping (Carrier 1993). While generosity might be idealized in the cliché 'it's the thought that counts', in practice much generosity at Christmas requires particular types of gifts: those bought new, at some expense, in the commercial economy. This particular social norm reinforces commercialization and reduces the possibility of more sustainable ways of celebrating Christmas (Ger and Belk 1999). Second-hand purchases, for example, are frequently viewed as 'unsuitable and offensive' Christmas presents (Tynan and McKechnie 2009, 248) – though this may be changing, and is contingent on the material thing in question (an antique clock or carefully purchased piece of vintage clothing being very different to a re-gifted item of superfluous interest – but mostly restricted to those with the means to buy antiques). Home-made gifts are likewise unsatisfactory to some people (Carrier 1993), although to others such items as hand-knitted clothing are considered special because of the emotion and labour invested in them. Consistent with other material objects in this book, use value – the embedded usefulness of a thing – remains another criterion for present giving and valuing.

The annual ritual of Christmas shopping is itself a trigger for mixed feelings of resentment and indulgence. Some value it almost as much as the unwrapping of gifts on Christmas morning, while others complain about the crowded malls, stress and weight of expectations. Through the very effort of Christmas shopping commodities are converted into gifts:

> Performing this ritual demonstrates that we can celebrate and recreate personal relations with the anonymous objects available to us... [we] can create a sphere of familial love in the face of a world of money... The sense of hard work, together with complaints about growing commercialization, help affirm that the impersonality of the commercial world does contradict family relations, and that people are, at

this heightened time of the year, really able to wrest family values from recalcitrant raw materials. (Carrier 1993, 63)

Related to this, gift wrapping – which like Christmas shopping is much-maligned in sustainability terms – brings about a partial 'transformation of meaning' between the personal and the commercial. Gift wrapping is an act of importance, transforming commodities into gifts (Cheal 1988).

'TIS THE SEASON TO BE JOLLY

Our ethnographic research suggests that there is no simple association between households that advance sustainability in daily routines throughout the year, and those that advance sustainability at Christmas. The households in our study who aligned themselves very closely to sustainability as an ideology, and enacted it in various forms of household practice throughout the year, were more likely to engage in (relative) excess at Christmas. Intriguingly, strongly 'green' households in our study were most likely to see their gift-giving at Christmas as relatively unencumbered by excess commercialism, even if items were purchased commercially. Sustainability might be occasionally advanced at Christmas in these households, and valued as such, but it was more likely to be an effect of managing the tensions between generosity and commercialism commonly felt in the general population, not a driver of either.

Indeed, the mass-produced items forming gifts in these households took on a significance which trumped sustainability concerns: as signs of generosity which, through rituals of shopping, wrapping and a sense of 'considered, thoughtful consumption' meant such items were perceived as different from similar items purchased for oneself, or acquired at another time of the year. For those who took a sustainability-as-frugality approach during the year, there was even less drive to be sustainable at Christmas: gifts need not be expensive, but an excess of small, commercially produced and individually wrapped items pointed to a need for abundance in order to celebrate family togetherness, a temper to the otherwise high frugality of everyday practices.

One participant, Evette, a married woman with two children living at home, celebrated Christmas on a tight budget in a way that was nevertheless fun and family-focused, with items like home-made gingerbread and beer. Evette was mindful of sustainability throughout the year, driven by both financial considerations and an expressed deep sense of care for the environment. She practised this concern by tending an organic fruit and vegetable garden and raising chickens. Her family of four shared bath water each night, water which was then reused for toilet flushing. Bicarbonate of soda and vinegar were used to clean everything in the home. Yet Evette's deep sense of environmental care did not appear prominently in her Christmas, which was focused on balancing financial concerns with a celebration of family life. Commercially acquired items were an important feature of their Christmas stocking traditions:

> Christmas Day is our gift time. The kids, even though Darryl's the youngest, he's 16, Vanessa's 24, they still put the stockings out... and I still go and get all stationery and stocking fillers at the [discount store] and always have some sort of junk ... so there'll be pads and rubbers and pens and all sorts of things in their stockings ... I've said to them, it's at that stage now that you've got everything that you could possibly need, you really do. I said, there's a limit now, it's not just free for all ... It was quite a generous limit originally, we used to just go all out. But with things the way they have been economically with the last couple of years, just our family, with Graham out of work, the limit's been coming down ... And the [children] are given the choice whether they want that in vouchers or whether they want that in gifts ... [yet] I try and put as many things wrapped as possible. So if they've got three pairs of undies I'll wrap the undies all separate so that it just sort of pads out the tree and makes it look ... it's opening the presents. And they still love it.

Lacking money to buy extravagant gifts was one issue, but it was a blow softened by numerous small, inexpensive and yet useful household items, like stationery and underwear, bought from the store and lovingly individually wrapped. Evette was unwilling to extend her concerns about sustainability to Christmas. She saw her everyday water-saving and cleaning practices as difficulties to be borne, albeit cheerfully, following the family's move out of a fancy home upon her husband's retrenchment. Such frugalities were not to be carried over into Christmas. Consumption of cheap items, 'junk' and wrapping paper, contribute significantly to the festivities. Christmas gift-wrapping was an important ritual, for transforming commercial transactions into things imbued with love.

For most of the households in our study, using elaborate rituals of shopping, wrapping and lighting, Christmas' material objects are imbued with particular social qualities that render them less encumbered by sustainability concerns than they might be if acquired through other means. In this way, although sustainability concerns could not be said to be entirely relinquished, neither were they embraced, and indeed took on a role alongside other concerns at Christmas, such as maintaining a household budget. Nevertheless, there remained a commonly felt dislike of Christmas-related excessive consumption, and its impact on sustainability. Such dislike was linked most closely to waste: food that was not eaten, gifts that were not needed or useful, and unnecessary packaging. Another householder in our ethnography, Marie, said:

> I hate the fact that [Christmas is] so over the top. I hate all the wasting, wasting food
> and packaging, everything and ... now that my children are older, I'm looking for
> presents that are not necessarily something that you gift wrap. I look for vouchers
> and things like that ... I've learnt my lesson with food because I just think people
> go so over the top with food and there's so much wastage ... And I just don't get it.

By choosing vouchers as gifts for her family, Marie felt she was able to significantly reduce her own contribution to excessive waste and packaging at
Christmas. In so doing, she left the choice of gift to the recipient, and thus,
implicitly, some of the responsibility for more sustainable consumption was
transferred.

Householders in our study frequently attempted to create boundaries
around 'too much stuff' at Christmas. Households sometimes instituted rules
about gifting practices, usually in negotiations with extended families.
Nevertheless, gift-giving was maintained as the core way in which the spirit of
Christmas – generosity – was expressed, and the value of family, and religion
in some cases, was affirmed. Gifts were persistently given, even when stipu-
lations were made by some family members that gifts should be cut back in
number, or reduced in financial value. Nobody wanted to reduce Christmas
gift-giving to nothing at all. Indeed, it was commonly thought that everyone at
a Christmas celebration must have at least one gift to unwrap. Furthermore,
even if households were relatively conscious of excess consumption, members
of their extended family were not necessarily so committed. And when it came
to a choice between family harmony and environment, family harmony would
win. Many things given at Christmas consequently piled up, filling up space
in homes, often unwanted as things. Yet they were also very welcome as signi-
fiers of family generosity.

WHAT TO DO?

There are some simple steps that can be taken to lessen the environmental
impact of Christmas: composting vegetable peelings; switching off Christmas
lights before going to bed; planning meals to avoid waste (Haq et al. 2007).
Recommendations from environmental groups regarding gifting practices tend
to focus on replacing material items as gifts with vouchers for services such as
massages and gardening or non-material ethical gifts, such as overseas aid
project sponsorships (ACF n.d.). They also recommend wrapping gifts in
newspaper or reused gift wrap. Such considerations are worthy from a sustainability perspective, but if interpreted as a need for frugality, environmental
considerations seem to threaten the deeply treasured rituals associated with
affirming family and generosity with material abundance. This is clearly

demonstrated in the significance attached to the values of conviviality, neighbourliness and festive pleasure associated with Christmas light displays in front yards. Togetherness and enjoyment – especially for children – are cited as important reasons for those producing these displays, people often of working class background, at least in England (Edensor and Millington 2009). Such reasons are much more important than concerns about good taste and design, limiting the possibility of reining in increasingly vibrant light displays. Middle-class observers, on the other hand, tend to express disgust and antipathy towards the waste and tastelessness in such displays. They are regarded by some as 'an immodest spectacle, excessive aesthetically, a waste of money by poor, inept consumers, and above all, a waste of electricity, an outrage against energy conservation and ecological thinking' (Edensor and Millington 2009, 109). Although perhaps specific to the class dynamics of England, and not indicative of a simple association between concerns for sustainability and class more broadly, it is instructive to note the conflicting role that Christmas lights play in marking Christmas as a festival of conviviality for some, and a sign of wasted resources for others. There seems to be no easy way to resolve this tension for Christmas traditions in general.

Similarly, one householder in our ethnographic study, Will, was adamant that he 'didn't want a goat for Christmas', an illustration of the deeply embedded assumption that Christmas giving is ultimately a material, and likely commercial, transaction between intimates, despite the rhetoric of a wider Christmas generosity. It is this wider generosity that charities attempt to harness, for social and environmental sustainability purposes, in the selling of items such as goats, chickens, beehives, and small business loans for communities in the global south, to people in wealthy countries looking for meaningful gifts. Will expressed what seems to be a common opinion that Christmas gifting practices are insufficient without personal receipt of a material thing, rendering mere rhetoric the value attached to giving (items such as goats) to those less fortunate. Thus, the moral position of generosity is rendered an individualistic and also a commercial one, despite laments about the over-commercialization of Christmas.

The wider challenge then is to decouple abundance and generosity from being underpinned by excessive commercialization and individualism, and to reconfigure them in other ways. This might include an emphasis on quality rather than quantity, and changes in Christmas rituals such as wrapping gifts with wrap that is permanent yet also beautiful. The thoughtful disposal of unwanted gifts could also become a ritual in itself like Christmas shopping, perhaps eventually helping to break down the taboo of second-hand gifts. It is worth remembering that Christmas is a set of rituals that have never been static. Its 'very old elements are… shuffled and reshuffled, others are introduced, original formulas perpetuate, transform or revive old customs' (Levi-Strauss 1993, 43). There is thus no reason to assume that the pursuit of sustainability, as a relatively new practice, cannot be incorporated more strongly into Christmas rituals over time.

18. Retirement

Retirement is a lifecycle transition with implications for sustainability as people change their work, consumption and often residential practices. The move out of the paid workforce has implications for household sustainability and the specific dilemmas discussed in other chapters. In this chapter we concentrate on three specific changes of the transition to retirement: in consumption patterns, dwellings and time.

The retiree population in most affluent Western societies is also in transition, from depression-era babies to baby-boomers. Current retirees who were born between the two world wars and retired at the end of the twentieth century often have a strong heritage of living frugally, and provide a benchmark of 'making do'. Those born after World War II – the 'baby-boomer' generation, now coming into retirement – have grown up in more affluent circumstances. They are also living longer, with expectations of an active retirement. We trace this second transition through the examples of frugality, driving and retirement migration.

The two transitions operate together to frame the sustainability dilemmas and environmental implications of an ageing population. An increasing population of healthy retirees provides a resource, for example through environmental volunteerism. There is also a set of specific environmental impacts, such as those associated with new migration patterns, a changing built environment (Nelson 2010) and increased health care requirements. An emerging research agenda around these issues (see Davies and James 2011, Pillemer et al. 2011 and Scarfo 2011 for recent reviews) indicates that much more work on the environmental impacts of ageing will need to be done over the coming decades.

RETIREMENT AS TRANSITION

Usually defined by the transition out of the paid workforce, retirement involves new patterns of everyday practice. Although some have argued that older people have a particular responsibility to live more sustainably for future generations, advocating for them to engage in 'legacy work' (Moody 2008), we do not assume that there is an increased ethical obligation among the elderly over any other age group. Further, it is important not to homogenize the ageing population (Davies 2011).

Consumption Patterns

Contrary to the predictions of economic models, consumption expenditure decreases substantially on retirement, even when that transition has been well planned financially (Lundberg et al. 2003; Davies and James 2011). In a US study, Lundberg et al. (2003, 1217) attributed a 9 per cent decrease in food expenditure in married households with a retiring husband to 'a model of marital bargaining in which wives prefer to save more than their husbands do to support an expected longer retirement period'. Luhrmann (2009) examined whether home production – in the form of cooking, cleaning, maintenance and repair – was substituted for consumer expenditure after retirement. In her German case study, she found a 17 per cent drop in expenditure and an increase in time spent on home production of 89 minutes per day. Specifically, 'retired households increase their home production mainly in the sector of cooking and preparing food, but they also engage more in doing washing, paperwork and gardening' (Luhrmann 2009, 241). These findings are consistent with New Zealand research showing a decreased per capita footprint (due to decreased resource use and waste production) in the 65+ age cohort, compared with those aged 15 to 64 (McDonald et al. 2006). When projected across the total population as it grows and ages, however, gains are diminished by the reducing proportion of young people (<15), who have the lowest per capita requirements.

Consumption patterns for retirees as much as anyone else are dictated by economic circumstances rather than age *per se*, as Stephens and Unayama (2012) show for Japan. There, significant retirement bonuses stimulate short-term increases in consumption for many households, but conversely, low-income households reduce their consumption at the point of retirement. Decreasing mobility and declining health can also contribute to reduced consumption. One elderly participant in our ethnography was almost entirely housebound in her modest public housing unit due to poor health. This factor, in addition to her reliance on a small government pension for income, meant that her consumption was lowest of the 16 households in our study.

Dwellings – Downsizing, Upsizing and Sharing

It is commonly accepted that retiree households, whether single or couple, are likely to 'downsize' to a smaller house or apartment once children have left home. This process can have a significant role in reducing greenhouse gas emissions, and the remodelling of the home provides opportunities for more sustainable interventions, as Doteuchi (2008) has argued for Japan. A team of students from our university is undertaking just such a challenge, via their entry in the Solar Decathlon: China 2013 competition (http://sbrc.uow.edu.au/sd/index.html). The team will retrofit a mid-twentieth century fibro-cement house as a more sustainable and flexible home for a retiring couple.

Flexibility is important. Households who can afford it may want to increase their house size at retirement, or improve it in other ways, because they will spend more time there (Gobillon and Wolff 2011). In our research into backyard gardens, we found that a surprising number of suburban grandparents in Australia chose not to downsize but rather to still structure their gardens around the needs of visiting grandchildren, for example keeping swings and cubby houses (Head and Muir 2007a, 46).

In other contexts elderly people move in with extended family members, such as adult children and their families. Increasing household size is one way to reduce the high overall ecological footprints associated with low numbers of people per dwelling. Nevertheless, as Klocker et al. (2012) demonstrate, such living arrangements require complex negotiations. They describe a phenomenon of *living together but apart*, in which the different generations create and maintain their own spaces – kitchens and bathrooms in particular – to maintain independence and privacy (see also Chapter 13). Decisions are not made in a straightforward way but are rather 'informed by competing desires to care for and support relatives, to maintain a sense of nuclear family, and an individualist predilection for privacy and space' (Klocker et al. 2012, 2256).

Time and Volunteering

A number of studies have demonstrated that having more time leads to high rates of volunteering among retirees, including in a variety of environmental activities (e.g. Davies and James 2011, 171), and that this also has good health and social outcomes for the people involved (Achenbaum 2008; Pillemer et al. 2010; Bushway et al. 2011). Indeed some argue that older people constitute 'a special resource for environmental action in the form of volunteerism and civic engagement' (Pillemer and Wagenet 2008, 1). In a preliminary US review Bushway et al. (2011, 195) found 'that older adults constitute a significant portion of the volunteer workforce of environmental organizations, and the potential exists for large-scale expansion of the numbers of participants especially in the context of declining fiscal resources'.

Warburton and Gooch (2007, 44) use the concept of 'generativity' to explain the desire of older people in particular to leave a positive legacy for future generations: '"Generativity" includes the desire of people to create something of lasting value, for themselves and for others; to care about others; and to feel valued themselves'. Analysing the experience of older people volunteering in environmental stewardship groups (for example, Bushcare and Landcare) in Queensland, Warburton and Gooch argue that such activities should be highly valued in the broader community, rather than governments seeing older people as a social burden. One of their interviewees explained his long-term perspective as follows:

> You feel as though you have to be in there for the long-term... even though there is immediate gratification... I'm thinking of hundreds of years down the track... I look at a tree which I know, has a lifespan of over 500 years, and I'll try and picture it... I try and wonder what sort of people will be around and if our birds and animals we are trying to keep from going locally extinct, will be still around... I can't think of anything which is more long-term. (Male, urban Bushcare group, quoted in Warburton and Gooch 2007, 48)

Nevertheless it is important not to generalize about older people having particular environmental views simply because of their age; other studies have found them to not be necessarily interested in environmental activism (Wright et al. 2003; see also Achenbaum 2008). Nor should older people be characterized as necessarily more concerned to leave an environmental legacy for the future. This is an empirically open question that needs to be explored in different cultural and geographical contexts.

RETIREMENT IN TRANSITION

Most current retirees grew up in more frugal times, but the coming generations of retirees will have had higher levels of affluence throughout their lives (Casado-Diaz 2006). In this section we show three examples of how retirement itself is in a process of change.

Cultures of Frugality and Making Do

Many of the people in our 2009 survey who identified their employment category as retiree/pensioner would have been born before 1945, although younger pensioners such as single parents would also be included. In this cohort we see the ways age and low socio-economic status interact to shape practices of frugality and thrift. Practices that we might think of as pro-sustainability behaviours – recycling and reducing electricity use – are often motivated by financial rather than environmental concerns. For example, retirees/pensioners were more likely than other employment status groups to 'always':

- save water in the bathroom by taking short showers;
- buy environmentally friendly detergents whenever possible;

- switch off lights in unoccupied rooms;
- buy products with as little packaging as possible;
- use their own bag when shopping rather than one provided;
- compost kitchen waste; and
- compost garden waste.

They were most likely (along with the unemployed) to 'always' repair clothing and to have a water-saving device fitted to their shower. They were also most likely (along with the self-employed) to use solar power, and to 'always' or 'usually' buy local produce wherever possible. In apparent contrast to the comments about always repairing clothing, retirees/pensioners were the least likely employment cohort to have repaired or reused something instead of throwing it away, in the three months prior to the survey (56.1 per cent – perhaps a stitch in time six months ago would have saved nine more recently). In that example, the most frugal were full-time students (89.7 per cent). The only other category in which retirees were least likely to undertake a pro-sustainability behaviour was that they were least likely to 'always' turn off the tap while cleaning teeth.

The high levels of composting referred to above are probably connected to the passion for gardening among many older people (see also Chapter 16). This has implications for sustainability in its potential for home-based food production. In our ethnographic work on backyards, retired Macedonian migrants with rural subsistence backgrounds had large gardens producing flowers, vegetables and culinary herbs. Food production in this example produced a variety of social goods, none of which were articulated in sustainability terms (Chapter 2).

In our research on extended family households there were complex generational differences in attitude and practice (Klocker et al. 2012). Younger generations – Gen Y – identified with sustainability by recycling, and by affirming their belief in the importance of tackling climate change. They thus claimed stronger 'green' credentials than their parents and grandparents. Yet it was grandparents who hated waste. They were least likely to consume large amounts of clothing and appliances, and instead kept and stored old stuff in many creative ways, maximizing its use value. In turn, grandparents in extended family households were most likely to distance themselves from 'green' identities and from the climate change issue. Our garden research has also demonstrated the strong frugality ethic among older Australians, and the declining power of an anti-waste mentality among younger generations (Head and Muir 2007a, 126). As baby-boomers and following generations age, it remains to be seen whether cultures of frugality will also decline.

Driving Behaviours

As one example of changing patterns, Rosenbloom (2001) used data from
the United States, Australia, Germany, New Zealand, Norway and the UK to
show that older people are already shifting from public transport to higher
car use than a decade previously. By 2020 there will be more than 48 million
drivers aged over 65 in the United States, and over 60 million in the EU
(Rosenbloom 2001, 402). Rosenbloom's research showed that in the United
States, for example, vehicle trip rates of older people did not reduce substantially until the age of 75. So retirees are not sitting quietly at home; they are
increasingly getting out and about by car. But they may also drive differently
from younger drivers. They undertake many short trips (which pollute
proportionately more because of the 'cold start' problem; see also Chapter
10). Up to 10 per cent of trips by older people are 'scouting' trips the day
before the real trip – to work out the best route, and locate parking facilities
for the next day (Rosenbloom 2001). Such patterns will require more
detailed research, but it is important that it not be undertaken in a way that
problematizes older drivers. Rather it is important to consider for society as
a whole how we can support the life aspirations of different age groups in
the most sustainably connected ways.

Retirement Migration – Associated Environmental Issues

Housing mobility at retirement is not only between particular dwellings, but
may involve migration to a new area, with attendant environmental impacts
(Gustafson 2001, 2008; Davies and James 2011; Gobillon and Wolff 2011).
Casado-Diaz (2006) attributed increased international amenity migration
among older people to a combination of increased life expectancy, the spread
of early retirement, rising incomes and affluence and the accumulation of
tourism experiences. On the north coast of Western Australia, Davies and
James (2011) argued that an increase in older amenity migrants led to
increased pressure on natural resources such as fish stocks, and increased
demand for housing, the latter leading to land clearing and increased pressure
on water supplies and waste disposal in the region. On the other hand, there
was also increased participation in community environmental activities.
Similar impacts have also been observed in the Alicante region of the
Mediterranean, where the traditional landscape is 'being replaced by a newly
emerging urbanized landscape, causing the marginalization of both the historic
cultural landscape and of natural elements, such as vegetation' (Zasada et al.
2010, 133). Demographic change unfurls slowly, yet has complex regional and
ecological consequences.

WHAT TO DO?

The concept of retirement – a period of relatively healthy life after stopping paid work – is itself a product of affluent societies. As more people enter the elderly cohort, with high expectations of and capacities for an active retirement, the sustainability implications become more complex. Both 'retirement' and 'the elderly' are categories that encompass considerable variability. As the examples in this chapter show, ethnicity, family connections and interests are all important. As with other issues, socio-economic status is to the fore; vulnerability and the capacity for choice fracture along socio-economic lines. Poorer people can substitute time for money in various household activities after retirement, exercising strong capacities for self-sufficiency, as long as they are physically able. Like other life transitions (Chapter 1), retirement has considerable potential for sustainable choices to be made in household patterns, but this cannot be assumed. Research and policy need to consider social and geographical context.

However, the older people get, the more they share vulnerabilities. The elderly are the most vulnerable to climate change impacts such as increased heat stress (Luber and Prudent 2009; Chapter 5). It is important to remember that ageing populations are also becoming more urbanized, and that many aspects of urban design will need to recognize these changes. Detailed discussions of climate change adaptations and the health implications of an ageing population are beyond the scope of this book, although see Farbotko and Waitt (2011) for our work on this issue. The point is that sustainability discussions – whether environmental or social or both – should not pit one age group against another. Instead we need to strengthen the connections between different groups in the community, while recognizing their specific needs and capacities.

19. Death

Unlike other issues throughout this book, death does not involve much choice. Nevertheless this does not mean that there are no choices to be made. Indeed those making decisions about the disposal of their body, or of a relative's body, are faced with a growing range of options if they wish to be 'green in death'. These are choices about the treatment of bodies prior to burial and where to place the body or the substance to which it has been reduced for burial. Associated activities such as flower-giving and transport to funerals are also significant. No different from other decisions or dilemmas discussed in this book, such choices involve the intersection of cultural, economic, and environmental issues. Indeed the deep and diverse cultural significance of death – of bodies, of graves, memorials, of beliefs about the afterlife or life after death, or indeed of one's legacy beyond life – makes this particular intersection striking for the dilemmas it engenders.

DEATH AND ENVIRONMENTALISM: IS DYING GREEN THE NEW 'DYING WELL'?

The disposal of the dead varies enormously over time and both between and within societies. What is constant is that bodies are never treated simply as waste (Davies 2005). Body disposal occurs through rituals that embody and demonstrate 'significant social values and...express how a people view the world and themselves', the nature of life and death, and the proper place and treatment of death among the living (Davies 2005, 55; Jalland 2006). This chapter focuses on death in Western cultures rather than attempt to cover the global diversity of practices and issues.

Secularization and the rise of the modern environmental movement have shifted norms. The contemporary influence of environmentalism on burial is not simply a pragmatic response to, for example, a shortage of burial land. More fundamentally, this influence represents a shift from a preoccupation with God, the afterlife, church authority and the eternal destiny of humanity, to a concern with the 'world as a living space, to ethical activity within it and to the likelihood of it having an ethical future' and to the rise of the notion of the spirituality rather than the religiosity of people (Davies 2005, 77). In other words, ecological framing of identity and the expression of this in death as well as in life is part of a broader trend to secular eschatologies (the doctrine of the final things – regarding one's state after death), less inclination to follow burial traditions for their own sake, and to more expression of individuality in funeral rites and burial.

As with other sustainability issues throughout this book, the rise in interest in 'green' burial may reflect sentiments of frugality rather than a concern with the environment *per se* (Clayden and Dixon 2007). A theme of several influential books on Western burial practices (Mitford 1998; Harris 2007; Larkins 2007) is that contemporary interest in 'green' burial practices is linked to resurgence in a persistent interest in more frugal, unadorned and family-managed burial practices – practices that arguably have been discouraged by a funeral industry intent on more elaborate and profitable burials.

There are a growing number of ways to 'green' funeral and burial practices (Harris 2007). These include reform of existing practices such as cemetery and crematorium efforts to become carbon-neutral (SMH 2008) or to use 'waste' heat to generate electricity (Copping 2011). Cremation is an option to reduce the impacts of managing a body and has a lower demand for land than burial. In addition there is a burgeoning range of alternative technologies and products to facilitate greener burials. The longest established alternative is probably natural burial, first 'revived' in the UK in 1993 (Clayden and Dixon 2007) and now found internationally. The forms of natural burial grounds vary. Essentially, bodies are buried in coffins of a readily degradable nature (cardboard or wicker) in a woodland or field setting, without memorial, and trees (usually native) are commonly planted on the gravesite. Other body disposal techniques that have emerged in response to sustainability concerns are alkaline hydrolysis in potassium hydroxide, pressure and raised temperatures (proprietary names include Resomation and Aquamation) and freeze-drying in liquid nitrogen, followed by sublimation and pulverization (proprietary names include Promession and Cryomation) (Everts 2010). The by-products of these processes can be interred as with cremation ashes, composted (freeze-drying), or applied as fertilizer (alkaline hydrolysis; although it may need pH adjustment). As well as these options are products designed to reduce the impact of one's death such as cardboard or recycled paper coffins and biodegradable shrouds and urns (the Natural Burial Company, for example, has an online store).

SUSTAINABILITY ISSUES

Death generates the problem of how to dispose of large amounts of organic matter sustainably – estimated to be 37 000 tons in the UK in 2007 (Monaghan

2009). With predicted global population growth from 6.9 billion in 2011 to 9.3 billion in 2050 (UN 2011) demands on resources by funeral and burial practices will only increase. For example, in Australia and the United States the number of deaths is increasing and this will drive continued demand for land for burial despite rising cremation rates (ABS 2010; Canning and Szmigin 2010). Urban land for burial has long been in short supply (Walter 2005; Rugg 2006). Competing uses are strong and suitable sites are limited. The potential conflicts are highlighted in a shortage of land for burial in Australia's largest city, Sydney (Marshall and Rounds 2011), where cemetery expansion plans not far from the central business district threaten a rare patch of commercial urban horticultural land (McKenny 2012). In the United States approximately 2.5 million people die each year and a comparable number of coffins are sold, consuming wood, metal, textiles, and finishing products such as varnishes (Harris 2007). In some countries and cemeteries burial also requires concrete vaults around the coffin to prevent collapse of the grave. Pollution is also a potential problem. Cemeteries are a 'particular kind of landfill, similar to unlined "dilute and disperse" municipal waste landfill sites' (Fiedler et al. 2012, 90). While consensus on cemetery pollution and its actual impact on human health is arguably lacking (Keijzer 2011), soil and water pollution is possible depending on the nature of the burial and the site, its soils, geology and climate (Dent et al. 2004). Pollution arises in part from the 'putrefactive liquid resulting from the body decomposition' which is mineral rich, has high biological oxygen demand, which can contain medical and 'chemical treatment residues', and pathogens associated with infectious diseases (Silva and Malagutti Filho in press). Specific constituent elements include nitrates, phosphates, faecal and other bacteria, and heavy metals (Zychowski 2012). Other potential pollutants include the products of coffin breakdown such as lead, iron, copper and zinc, polychlorinated biphenyls, methanol, and solvents.

Alternatives to burial are not without their problems. High temperature cremation, the main alternative to burial in the West, has at least two significant problems. First, cremation is a potentially significant source of mercury (and other harmful substances) due to its presence in dental amalgam. In the UK and other European countries, many crematoriums are updating their furnaces to meet new mercury emission requirements (Copping 2011). Second, and not least, are the energy consumption and the carbon emissions associated with cremation. Published research on the topic is elusive but a number of private and media reports suggest that resource impacts can be substantial. In Australia cremation occurs in furnaces heated to between 760 and 1150 degrees Celsius, takes between 70 and 150 minutes, and, for an average Australian male, uses 160 kg of gas (ERC 2011; Luntz 2007). Estimates of carbon emissions per cremation vary from two Australian figures of 50 kg

(Luntz 2007) and 160 kg (Parliament of South Australia 2008; 120 tonnes for the state in 2006/7, about the same emissions as running 267 cars for a year) to 100 kg in a Dutch analysis (Keijzer and Kok 2011).

We also need to consider the overall impact of burial practices, and potential contradictions between professed commitment to sustainability and actual outcomes. An example from Harris' (2007) book on 'green' alternatives to mainstream burials highlights this. At its site in Florida, Eternal Reefs will incorporate cremated remains into concrete 'reef balls' which are sunk into the ocean to rehabilitate or create reefs and thus a 'permanent environmental living legacy' (Eternal Reefs n.d.). This constructive option buries divers, fishermen and the like in an environment important to them, avoids land burial, and presumably provides benefits to the marine environment. However, Harris (2007) describes how relatives fly in to Florida from other US states for the manufacture of the reef balls, which are then transported interstate to the chosen sinking site if desired. Under these conditions (and concrete use aside), in terms of overall sustainability, the reef ball option is going to have a high environmental impact. This raises the need to look beyond 'single' issues such as land or pollution to the broader impacts of different ways of dealing with the dead.

A recent Dutch analysis of 'standard' (i.e. not ornate or simple) funeral and burials (Keijzer 2011) used lifecycle assessment to analyse the broader environmental costs of different burial options. The results were complex. Burial had the highest environmental impacts largely due to land requirements, mechanized grave excavation, monuments, and the fuel use and emissions resulting from cemetery maintenance. Cremation followed at generally half the impacts of burial. Cremation impacts were largely due to gas use and flue gas emissions such as carbon dioxide, methane, and nitrogen oxide. The two newer burial techniques, alkaline hydrolysis of the body and freeze-drying, had the lowest impacts and were comparable to each other. Their impacts were largely due to direct and indirect electricity use and to assisting processes such as car transport.

A key finding of this analysis was that the bulk of the impacts were not generated by body disposal itself but by other activities associated with burials. Those activities that contributed the most were correspondence such as cards, flowers, food and beverages (such as at a wake), and transport to and from the funeral service or burial. Even excluding a range of activities and products for which there was ambiguity or insufficient information, such as body preparation and clothing, embalming, the funeral ceremony itself, and even coffin parts such as handles, these associated activities had about three times the environmental impact of burial, the highest impact form of body disposal (Keijzer 2011).

TRANSFORMATION AND COMPLEXITIES

In a consumer-oriented analysis of death and burial, Canning and Szmigin (2010, 1132) argued that 'acknowledging the impact that the nature of disposal has on the maintenance of identity beyond death is critical to understanding the cultural and social imperatives felt by many in a constrained place of decision-making'. This is relevant to the deceased and any choices they might have made about their burial as well as to family who may wish to recognize the identity of the deceased, meet church norms, or fulfil their own needs such as having a memorial to visit (Kellaher et al. 2005). Any imperative to change practice may not be readily and universally accepted. On the other hand, the variability of disposal practices and their connection to changing social values, suggests openings for innovations in body disposal that are likely to enhance the sustainability of death.

Anyone wishing to have or organize a sustainable burial will need to navigate a range of social and institutional realms that may frustrate or facilitate their wishes. These realms may include the funeral industry, the national legal and institutional setting, and a wide range of cultural preferences that may influence groups, families or individuals. A few examples illustrate this. First, death is something that most people only deal with occasionally. If preparations for a death have not been made in advance, people are likely to have to make quick decisions under distress; shopping around or exploring options may be difficult or 'taboo' (Canning and Szmigin 2010). They remain dependent on a funeral director and their particular offerings. While 'green' burial options are becoming more common in the industry, this is not universal and there may be no time to assess 'green' options against their claims. Further, going out on what may be a 'green limb' at such a time may not be an option that bereaved family members are willing to choose. Second, national 'burial cultures' may vary such as in the distinction between the 'celebratory and costly' American model and the 'more utilitarian' approach in the UK (Canning and Szmigin 2010, p. 1131). In the American model, expensive and ornate funerals and burials have been a means of expressing status and the regard in which the deceased is held (Mitford 1998). Third, national institutional frameworks can also frustrate burial alternatives. The 'municipalization of disposal' in Europe has been identified as promoting cremation but also as limiting disposal of ashes (Walter 2005). Scattering is commonplace in Britain but Germans must bury ashes in approved cemeteries – with the consequence that Germans scatter ashes in Switzerland, causing concerns there about impacts on a reservoir near one popular scattering location (Canning and Szmigin 2010).

Fourth, people may have to navigate and weigh up a range of social and cultural preferences and beliefs among the groups to which they belong. European migrants in Australia opposed cremation and were concerned about the sanctity of the body until resurrection; they increasingly demanded mausoleums (Jalland 2006). Australian authorities who preferred cremation as a way to deal with space, maintenance, and environmental issues nonetheless allowed mausoleum development, consumptive of land, materials, and maintenance resources. Among Jews, cremation is forbidden and burial, consumptive of land, is the norm, although burials are traditionally simple, with a plain wooden coffin (Levine 1997). As a final example, cultural preferences may complicate the spread of natural burial as a 'greener' option in some countries. In the UK trees are planted as part of natural burial act – a permanent memorial for family and friends – while such living memorials are regarded as neglectful in some European countries where fresh flowers on a grave signify fulfilment of duty to kin and ongoing relationships between the living and the dead (Clayden and Dixon 2007).

We also need to consider the dilemmas raised by Keijzer (2011) with her argument that the highest impact of funerals and burials stems from practices such as sending flowers, transportation, and providing food and beverages at wakes. These are critical issues, for it is through such practices that respect and love for the dead, friendship and familial bonds, support, and values are enacted, affirmed, and offered (Howarth 2007). Consider the reactions if one announced that one was not attending a family funeral on the grounds of the fossil fuels consumed and carbon emitted in travelling to it. This is not yet within the bounds of acceptable excuses.

Finally, cost is likely to be a factor in decisions about whether to pursue a 'green' burial of some sort. Costs vary enormously but it seems likely that 'green' options are not necessarily cheaper. Gravesites in natural burial cemeteries can be comparable in cost to conventional grave sites (Clayden and Dixon 2007; The Centre for Natural Burial 2010; Potts 2011). Cardboard coffins are a more sustainable option but, according to one manufacturer, are still required to meet standards for coffin construction, and so are priced within the 'range of traditional value coffins' (LifeArt Australia n.d.). Cryomation has been reported as being similar in cost to cremation in Sweden (Natural Ways 2007) and Aquamation Industries in Australia expects their form of alkaline hydrolysis to cost consumers a similar amount to cremation (John Humphries, Aquamation Industries, personal communication 2012). Nevertheless the possibility exists for the market to corner 'green' options into a pricier niche, mirroring other forms of consumption (see also Chapters 3, 15).

WHAT TO DO?

In death, as in life, people make choices that reconcile often patchy information with their own values, often at a time when decisions need to be made quickly. In some cases, such as the limited availability of hydrolysis and freeze-drying, access to 'green' options may impose limits on what can be done. It may be that, without travelling, the nearest and 'greenest' option is cremation. There are a number of issues that families will have to negotiate among themselves to find acceptable burial outcomes; it is likely that even a decision to have some sort of 'green' burial will involve compromises with what is socially or culturally acceptable.

That said, as a consumer of a service and associated products, what can one do to limit the environmental impact of one's death? First, one can plan ahead to explore options and rights, and talk to loved ones (as with organ donation). In your area can you be buried in your garden? Can you keep the body at home and transport it yourself in your station wagon to the crematorium? Is there a natural burial ground nearby? Is a coffin even required? These can all reduce dependence on the standard offerings of local funeral directors as well as lower impacts by minimizing transport of the body and the need to buy goods. This may involve discussing what you want with funeral directors and seeking their assistance in organizing it. Alternatively, seek out one of the many not-for-profit organizations, including many who specialize in 'greening' burial, who can provide advice on planning funerals and burials. Through such organizations, it may, for example, be possible to identify 'green' options in your area, find suppliers of biodegradable coffins or urns that can be purchased in advance, or find body disposal facilities or funeral directors willing to help people be more sustainable in death.

Second, and linked to many of these issues, keeping funerals and burials simple appears to be a key factor in minimizing the environmental impact of death. This might mean choosing a simple coffin without metal fittings and unnecessary linings, minimizing treatment of the body, minimizing flowers, and providing food at wakes from sustainable sources. Transport of mourners is harder for the bereaved to control. Keijzer (2011) calculates such actions could reduce impacts up to 59 per cent. Similar steps can be undertaken for other elements of burial and funerals. Keijzer (2011) calculates that not having a stone grave monument can reduce the impacts of a standard burial by a third and a cardboard coffin could reduce impact by 8 per cent.

None of these options necessarily involve any less care of the dead. Indeed those who support simpler burial practices, reclaiming death and bringing it closer to the living, would see opportunities rather than constraints in the above suggestions for families to care more closely for their dead. The challenge to consumption is more profound. While many of these suggestions involve a reorientation of material consumption and generate economic opportunities, there remains a clear challenge to an industry with much invested in selling ornate coffins or with capital invested in facilities such as crematoria.

20. Conclusion

From Kyoto to Copenhagen, Durban to Rio, international groupings of nation states have failed to deliver the climate change action that was widely hoped for, and for many politicians the issue has slipped into the 'too hard' basket for the time being. Nevertheless, as the writing of this book comes to a close, scientists have renewed concerns for urgent action. In 2012 Arctic sea ice melted to a record low, and the northern hemisphere experienced its hottest summer yet – breaking 3215 high-temperature records in June, in the United States alone (McKibben 2012). Scientists suggest key biophysical tipping points – irreversible melt of the Greenland ice sheet, resulting in leaking of billions of tonnes of methane currently trapped in that ice; equatorial moisture build-up, 'loading the dice' for more intense monsoons, hurricanes and devastating floods – will likely be passed somewhere between the years 2016 and 2030, well within the readable lifetime of this book. The Australian Government Climate Commission (2012) argues that the present decade – to 2020 – is critical if humans are to prevent logarithmic increases in CO_2 emissions and global warming.

Governments have introduced incentives for 'green' technological innovation, improved ecolabelling schemes and the like. Yet regulators have continued to open up new fossil fuel deposits for exploitation and there remain enormous (and growing) reserves of oil, coal and gas already factored into the asset sheets and investments of the world's largest mining and oil corporations. The 'distinction between knowing and doing' (Davidson 2012, 112) persists: we know that climate change requires substantial action, but we continue to do unsustainable things.

To return to a question we posed in the introduction to this book, is the household one place to imagine meaningful change?

One clear conclusion from this book is that any changes that a household makes are limited unless connected to larger-scale movements. Boundless consumption remains ever-possible for the wealthy, inviting the ready exercise of scepticism by the less well off. Options to consume remain largely open – freedom and choice are themselves frictions of sorts. Immediate issues are whether ceaseless economic growth, Western cultural norms of cleanliness, comfort and opulence come to be the aspirations for industrializing countries of the global south (which were largely beyond the scope of this book), and what that might mean for the future of the planet. Improved sanitation is a critical justice issue, potentially improving millions of lives. Does it also require widespread adoption of Western technologies and norms of toileting? Will car and television ownership patterns in North America be replicated globally? What happens if, as some futurists predict (Glance 2011), 100 billion things worldwide – from cars and cows to cereal boxes – become connected to electrical flows via radio frequency identification tags, screens and GPS chips? Even if nuanced examination of affluent households could be harnessed to transform everyday practices that reduce CO_2 emissions, gains made could well be eclipsed by capitalism's endless thirst for profits (and data), ratcheting connectivity to electrical grids and aspirational consumerism elsewhere.

Throughout this book we have seen that one-dimensional conceptions of households do not match with their actual complexities, contrasts and contradictions. Households are unpredictable. People are stubbornly fixated on some things – such as what it means to be 'clean' (Chapter 7), the need for automobility (Chapter 10), and for privacy within household spaces when watching television (Chapter 13) or urinating or defecating (Chapter 6) – and yet are not necessarily so stubborn about others, such as water frugality (Chapter 4) or heating and cooling practices (Chapter 5). Households are not uniform or passive, and are unlikely to 'play ball' when governments or corporations might wish them to.

There are, in short, important questions of lifestyle, comfort and cultural norms that plug into the bigger picture. These are not small things that might prove useful in, say, a century, but rather the mundane, and yet utterly pervasive, everyday decisions, practices and behaviours that pattern human life. The microscopic details of household life matter enormously: they are the cellular activity fuelling climate change; they are the lifestyles, relationships and emotions that most people care and worry about.

Another clear conclusion from the book is that households are not homogeneous, socially or geographically. They are nuclear families within which parents argue with teenagers about leaving lights or heaters on; they are baby boomers approaching retirement who argue over which stuff to keep and which to throw out; they are extended family households who fight over the television remote control, but who also have impressive capacities for sharing and thus reducing per capita energy and resource use. They are single-person households, student households, couple households in old age, families scraping to get by, blended families, same-sex couples with and without kids, living in homes in densely populated cities, in coastal towns, in global cities and downtrodden districts. Nowhere do households consume stuff or approach environmental issues in identical or predictable ways. And yet we do need to transcend our differences towards collective action, if we are to make a difference in our lifetime.

Contradictions abounded across the survey and ethnographic data on which this book was built. The wealthiest bracket of households were twice as likely to install solar power (although still in very small numbers) as the poorest, but were also the most prevalent users of air conditioning. The poorest households were most likely to say that they were 'uninterested' in climate change as an issue, but they were also the least likely to own LCD or plasma screen televisions or clothes dryers. The poorest households were also the most likely to repair clothing; to use toilet paper made from recycled paper; to buy 'environmentally friendly' detergents; to reuse glass bottles and jars; and to save water by taking shorter showers. Those with higher levels of educational qualifications were equally likely to use air conditioning habitually during summer as were those with basic high school education. Baby boomers were the least likely to be sceptical about climate change, but the most likely to fly five times or more annually. Such contradictions repeated themselves across myriad types of consumption within households. Practice, identity and attitudes do not necessarily line up.

Some households were conscious of their choice in purchasing decisions; others cared very little about it – but neither should be seen as necessarily 'good' or 'bad' from a sustainability perspective. In the case of both clothes (Chapter 3) and mattresses (Chapter 8) 'aware' consumers were more likely to purchase 'eco' products, and more likely to pay for recycling materials, to buy second-hand or to donate to charity shops, but were also the most likely to buy more clothes, stockpiling more than they need, and to replace their mattresses more often, thus fuelling greater levels of global resource use. Meanwhile the 'apathetic' were the most likely to buy cheap, poorly made clothes and mattresses, those with questionable environmental imprints, and least likely to recycle, give away to charity shops or to buy second-hand. Nevertheless on the whole they consumed less, and extracted maximum use value out of purchased items, extending lifespan. Contradiction confounds linear thinking about problems and solutions.

A third conclusion is that enormous knowledge and capacities exist. In households where frugality is a necessity rather than choice, creativity and adaptability is needed to make ends meet. Families find ways to achieve quality of life without stockpiling material things, without air conditioners or SUVs. Everywhere – from clothes to televisions, furniture to Christmas presents – people continue to privilege the usefulness of things. They keep, store and gift because usefulness still resonates as a core human value. This is an enormous potential source of traction. People are still alive who grew their own food or mended clothes during periods of wartime rations – a reminder that there are effective systems of provision beyond the industrial capitalist system and stocks of knowledge not yet lost. Migrant families bring with them other ways of knowing and relating to non-human nature (Klocker and Head,

2013). There are ways to be generous and to celebrate abundance, such as around Christmas, without having to sacrifice expressions of love for family and friends (Chapter 17).

There are also differences in what is possible at home versus as work or in public. People were much more willing to feel sweaty or dirty at home, where a sense of 'togetherness' may be fostered (Chapter 7). Thermal variation was tolerated at home, where beanies and slippers could be worn without embarrassment, more than in offices, where 'all-season' business suits remain the norm (Chapters 3, 5). It may be easier to get traction on a range of fronts in domestic than in business contexts.

Fourth, there are perplexing dilemmas of household sustainability, some technical, some socio-economic, some cultural. It is not so easy being 'green' – although low-income households and women are already doing much of the work of sustainability without grandstanding it. Throughout we have sought to suggest practical things that can be done, taking into account known science and cultural constraints. Sometimes the dilemmas are irresolvable: the eReader versus books, the great nappy debate, whether the gains in efficiency in high-density living outweigh the benefits of gardens in larger suburban blocks. At what point are the gains made on keeping old appliances offset by their energy intensity – when newer, much more efficient models are available? This tension can become a friction or a source of traction depending on decisions that householders make, one way or another – or because of the advice given by gatekeepers such as mechanics, installers or interior designers (Chapters 12, 15). Many dilemmas require the disentangling of cultural practices from wider structural imperatives; others require behavioural change to work in tandem with regulation. We need to reduce driving and flying, but that means we also need to build our cities and structure our business practices and holidays differently. We need to wash clothes and flush toilets less often, but that means viewing odour, waste and dirt differently.

Fifth, beyond creativity and technological innovation, we need a stronger sense of human stewardship over and interaction with material things (Lane and Watson 2012). We do not have much time to ponder the nuances, but rethinking our entanglements with non-human others – within and beyond the home – seems essential. Viewing the material things around the home as inert, or the non-human living things in our homes and gardens as without power, merely 'feeds human hubris and our earth-destroying fantasies of conquest and consumption' (Bennett 2010, ix). The assumption of intrinsically inanimate matter 'may be one of the impediments to the emergence of more ecological and more materially sustainable modes of production and consumption' (Bennett 2010, ix). This might mean, among other things, rejecting monotony or the impulse to immediately discard; going with the flow, appreciating seasonal rhythms in food production and indoor temperature; working together to save water in drought, but also savouring abundance when it is there; wrestling with time constraints around work, family and food, but tolerating late trains and finding time to walk, travel slowly and talk to others.

Indeed, the choices we make about material things, and the relationships we build with humans and non-humans alike, reveal our deeper self (Hawkins 2006). Do we envisage ourselves as the independent, autonomous, rational subject that relates to the consumption and disposal of things instrumentally in economic terms – cheap or fast is best, disposal is someone else's problem? Alternatively, do we envisage ourselves in relational terms, where our responsibility to others is understood as a way of caring for the self? Perhaps this is the greatest challenge of all: to begin to understand the self as somehow morally responsible not just for itself, but for others, human and non-human.

Stewardship suggests the need to focus on the utility and intimate history of material things. Our household furniture currently connects us to labour exploitation and deforestation in Asia (Chapter 8), our air conditioner to coal-fired power stations (Chapter 5), our own excrement to ocean life and landfill well out of sight (and smell) (Chapter 6). In lots of cases those environmental traces are not at all clear to consumers – but in many instances, too, we continue to consume things (like cheap clothing made in the global south) knowing full well that their production likely involved exploitative labour practices and environmental damage. Stewardship might also mean knowing when to recycle or pull apart something old and yet still functional, when it crosses the threshold and creates other kinds of ongoing environmental burdens. Retrofitting homes and lives requires willingness to alter things when they become problems, no matter how habitual.

Throughout this book were also counter-examples where households developed more constructive and reflective modes of interaction and stewardship – clothes swapping, handing down furniture, opening windows to catch the late afternoon breeze, putting up with the smell of family urine so that toilets were flushed less often. As Lane and Watson (2012) admit, there are limits. Immanent in the basic concept of stewardship are dispositions towards consumption and property ownership at odds with the axioms of the global market economy. But within and beyond the home are abundant examples of how people share space, resources, appliances, food, books and much more outside the cash economy. Such acts might not yet be enough to overhaul the economy, or extend to all areas of consumption, but they are a resource to be valued.

A related question is who is *responsible* for the environmental imprint of stuff, including its disposal and recycling. The assumption that this rests on consumers is fraught with ethical and practical problems. The paradox between manufacturer competitiveness and environmental goals is not universal, but it nevertheless continues to plague such industries as white goods

(Chapter 12), electronics (Chapter 13) and mobile phones (Chapter 14). Successful product designs based on features rather than energy efficiency are 'locked-in' and then only minimal change is possible around the margins. As Handfield et al. (1997, 311) argued, 'design engineers must consider alternative processes in the design function… They must also have objective, realizable goals in terms of waste reduction, not pie-in-the-sky goals such as "zero emissions" and other clichés'. One such goal may be to prioritize upgradeability and modularity in design – lessening the need for wholesale replacement of such things as phones, laptops and fridges.

Meanwhile, as the cases of televisions and mattresses show, the private sector frequently fails to take up the responsibility of waste reduction. Given low market prices for recycled goods the burden instead falls on households to do the right thing, or on non-profit and charity organizations to do the technical 'grunt work' of recycling vast quantities of materials. A reinvigorated role for government is the only long-term answer. Where the state has come to impose stronger regulations on manufacturers, and made them responsible for stewardship, change has started to catalyse. This will in turn mean, at the very least, that right-wing axioms of 'free market' production, without regulation, are unsupportable on sustainability grounds.

What lies ahead is, therefore, much more than a gradual adoption of 'green' technologies in the home while we wait for grander international solutions. Such solutions may never come. Rapid global warming and ecological volatility appears upon us, and we can only expect households to do so much. The challenge is nothing less than immense – the need to coordinate, quickly, promotion of sustainable consumption, collectivization of efforts (underpinned by technical and social research), profound transformation in government regulation and in the economy itself, 'the co-evolution of domestic practices, systems of provision, supply chains and production' (Røpke 2009, 2490). We hope here to have gone some way to provide insights into one element of all this: the complexities and nuances of household life, and the material things found in the home. Everyday life is characterized by diversity and texture, capacity and adaptability. In the event that neither systems of production and consumption nor cultural habits and norms shift in time, such capacities and adaptabilities of households will likely prove vital to our survival and possible future happiness.

References

Aalbers, M. (2008), 'The financialization of home and the mortgage market crisis', *Competition and Change*, **12**, 148–66.

ABC (Australian Broadcasting Corporation) (2012), 'Deadly cold snap cripples Europe', 6 February, accessed 6 February www.abc.net.au/news/2012-02-06/cold-snap-brings-europe-to-a-standstill/3812704.

ABS (Australian Bureau of Statistics) (1999), *National Nutrition Survey, Foods Eaten Australia, 1995*, catalogue no. 4804.0, Canberra: ABS.

ABS (2003), *Yearbook Australia, 2003*, catalogue no. 1301.0, Canberra: ABS.

ABS (2006), *Environmental Issues: People's Views and Practices, Mar 2006*, catalogue no. 4602.0, Canberra: ABS.

ABS (2008a), *Environmental Issues: Energy Use and Conservation*, Canberra: ABS.

ABS (2008b), *Environment and Energy News, Dec 2008*, Canberra: ABS.

ABS (2009), *Are Households Using Renewable Energy?*, catalogue no. 4102.0 Australian Social Trends, March, Canberra: ABS.

ABS (2010), *Burial and Cremation Trends in South Australia*, Canberra: ABS.

ABS (2011a), *Household Use of Information Technology, 2010–11*, Canberra: ABS.

ABS (2011b), *Environmental Issues: Energy Use and Conservation: March 2011*, Canberra: ABS.

Accenture (2012), 'Always on, always connected', accessed 4 September 2012 at www.accenture.com/SiteCollectionDocuments/PDF/Accenture_EHT_Research_2012_Consumer_Technology_Report.pdf#zoom=50.

ACF (Australian Conservation Foundation) (2007), 'Consuming Australia', Melbourne: ACF, accessed 17 July 2012 at www.acfonline.org.au.

ACF (no date), 'The hidden cost of Christmas: the environmental impact of Australian Christmas spending', Melbourne: ACF, accessed 24 June 2012 at http://acfonline.org.au/uploads/res/res_xmascost.pdf.

Achenbaum, W.A. (2008), 'From "green old age" to "green seniors": a synoptic history of elders and environmentalism', *Public Policy & Aging Report*, **18** (2), 8–13.

Ackermann, M. (2002), *Cool Comfort: America's Romance with Air-Conditioning*, Washington, DC: Smithsonian Institution Press.

Adey, P., L. Budd and P. Hubbard (2007), 'Flying lessons: exploring the social and cultural geographies of global air travel', *Progress in Human Geography*, **31**, 733–91.

Albanese, A. (2011), 'Sydney needs a second airport', press conference, 5 April, Federal Minister for Transport and Infrastructure, accessed 11 May 2012 at www.minister.infrastructure.gov.au/aa/pressconf/2011/APC018_2011.aspx.

Aldaya, M.M. and A.Y. Hoekstra (2010), 'The water needed for Italians to eat pasta and pizza', *Agricultural Systems*, **103**, 351–60.

Allon, F. and Z. Sofoulis (2006), 'Everyday water: cultures in transition', *Australian Geographer*, **37**, 45–55.

Allwood, J.M., S.E. Laursen, C.M. Rodriguez and N.M.P. Bocken (2005), *Well Dressed? The Present and Future Sustainability of Clothing and Textiles in the United Kingdom*, Cambridge: Cambridge University Press.

Ampt, P. and K. Owen (2008), *Consumer Attitudes to Kangaroo Meat Products*, Canberra: Rural Industries Research and Development Corporation.

AMTA (Australian Mobile Telecommunications Association)(2011), 'MobileMuster 2010–11 Annual Report', accessed 13 August 2012 at www.mobilemuster.com.au/news/annual-reports-publications/.

Appleby, B. (2010), *Skippy the 'Green' Kangaroo: Identifying Resistances to Eating Kangaroo in the Home in a Context of Climate Change*, BSc (Hons) thesis, University of Wollongong.

Arslan, B.A.E., S.W. Boyd, W.K. Dolci, E.M. Dodson, S.C. Boldt and B. Pilcher (2011), 'Workshops without walls: broadening access to science around the world', *PLos Biology*, **9** (8), 1–4.

Ashenburg, K. (2007), *The Dirt on Clean: An Unsanitized History*, New York: North Point Press.

Asif, M. and T. Muneer (2006), 'Life cycle assessment of built-in-storage solar water heaters in Pakistan', *Building Services Engineering Research and Technology*, **27** (1), 63–9.

Askew, L.E. and P.M. McGuirk (2004), 'Watering the suburbs: distinction, conformity and the suburban garden', *Australian Geographer*, **35**, 17–37.

Aumônier, S., M. Collins and P. Garrett (2008), *An Updated Lifecycle Assessment Study for Disposable and Reusable Nappies*, Science Report SC010018/SR2, Bristol: Environment Agency.

Avvannavar, S.M. and M. Mani (2008), 'A conceptual model of people's approach to sanitation', *Science of the Total Environment*, **390**, 1–12.

Bakar, W.A. (2010), *Measuring and Exploring Perspectives on Food Insecurity*, PhD thesis, University of Wollongong.

Barnes, C. (2008), 'Solar hot water systems buying guide', Choice Australia, accessed 25 May 2012 at www.choice.com.au/reviews-and-tests/household/energy-and-water/solar/solar-hot-water-systems.aspx.

Barr, S. (2003), 'Strategies for sustainability: citizens and responsible environmental behaviour', *Area*, **35**, 227–40.

Barr, S. (2007), 'Factors influencing environmental attitudes and behaviors: a U.K. case study of household waste management', *Environment and Behaviour*, **39**, 435–73.

Barr, S. and J. Prillwitz (2012), 'Green travellers? Exploring the spatial context of sustainable mobility styles', *Applied Geography*, **32**, 798–809.

Barr, S., G. Shaw, T. Coles and J. Prillwitz (2010), '"A holiday is a holiday": practicing sustainability, home and away', *Journal of Transport Geography*, **18**, 474–81.

Basker, E. (2005), 'Twas four weeks before Christmas: retail sales and the length of the Christmas shopping season', *Economic Letters*, **89**, 317–22.

Baxter J., B. Hewitt and M. Western (2003), *Post Familial Families and the Domestic Division of Labour: A View From Australia*, School of Science, University of Queensland.

BBC (British Broadcasting Corporation) (2011), 'Carbon footprint of watching television', BBC, accessed 4 September 2011 at www.bbc.co.uk/blogs/researchanddevelopment/2011/10/tv-carbon-footprint.shtml.

Beard, N.D. (2008), 'Branding of ethical fashion and the consumer: a luxury niche or mass-market reality?', *Fashion Theory*, **12**, 447–68.

Beder, S. (2012), 'The real cause of electricity price rises in NSW', 3 September, accessed 18 September 2012 at https://theconversation.edu.au/the-real-cause-of-electricity-price-rises-in-nsw-8955.

BedTimes (2009), 'Research finds 5 key mattress consumer segments', January, available online, accessed 16 June 2012 at http://bedtimesmagazine.com.

Belk, R. (2010), 'Sharing', *Journal of Consumer Research*, **36**, 715–34.

Belobaba, P. and A. Odoni (2009), 'Introduction and overview', in P. Belobaba, A. Odoni and C. Barnhart (eds), *The Global Airline Industry*, Chichester: Wiley, pp. 1–17.

Bennett, J. (2010), *Vibrant Matter: A Political Ecology of Things*, Durham, NC: Duke University Press.

Bennett, T., M. Emmison and J. Frow (1999), *Accounting for Tastes*, Cambridge: Cambridge University Press.

Berners-Lee, M. (2010), *How Bad are Bananas? The Carbon Footprint of Nearly Everything*, London: Profile Books.

Berners-Lee, M., C. Hoolohan, H. Cammack and C.N. Hewitt (2012), 'The relative greenhouse gas impacts of realistic dietary choices', *Energy Policy*, **43**, 184–90.

Berry, T. and J. Goodman (2006), 'Earth calling … the environmental impacts of the mobile telecommunication industry', Forum for the Future, accessed 11 September 2012 at www.forumforthefuture.org/sites/default/files/.../earthcalling.pdf.

Biehler, D.D. and G.L. Simon (2011), 'The great indoors: research frontiers on indoor environments as active political-ecological spaces', *Progress in Human Geography*, **35**, 172–92.

Bissell, D. (2007), 'Animating suspension: waiting for mobilities', *Mobilities*, **2**, 277–98.

Black, D.A., J.A. Black, T. Issarayangyun and S.E. Samuels (2007), 'Aircraft noise exposure and residents' stress and hypertension: a public health perspective for airport environmental management', *Journal of Air Transport Management*, **13**, 264–76.

Black, M. and B. Fawcett (2008), *The Last Taboo: Opening the Door on the Global Sanitation Crisis*, London: Earthscan.

Blunt, A. (2005), 'Cultural geography: cultural geographies of home', *Progress in Human Geography*, **29**, 505–15.

Blunt, A. and R. Dowling (2006), *Home*, London: Routledge.

Böhm, S., C. Jones, C. Land and M. Paterson (2006), 'Part one conceptualizing automobility: introduction: impossibilities of automobility', *The Sociological Review*, **54**, 1–16.

Boström, M. and M. Klintman (2011), *Eco-Standards, Product Labelling and Green Consumerism*, Basingstoke: Palgrave Macmillan.

Bows, A. (2010), 'Aviation and climate change: confronting the challenge', *The Aeronautical Journal*, **144** (1158), 459–68.

Bows, A., K. Anderson and S. Mander (2009), 'Aviation in turbulent times', *Technology Analysis & Strategic Management*, **21**, 17–37.

Bows, A., P. Upham and K. Anderson (2005), *Growth Scenarios for EU and UK Aviation: Contradictions with Climate Policy*, Manchester: Tyndall Centre for Climate Change Research.

Boynton-Jarrett, R., T.N. Thomas, K.E. Peterson, J. Wiecha, A.M. Sobol, S.L. Gortmaker (2003), 'Impact of television viewing patterns on fruit and vegetable consumption among adolescents', *Pediatrics*, **112**, 1321–6.

Bradshaw, A. and R. Canniford (2010), 'Excremental theory development', *Journal of Consumer Behaviour*, **9**,102–12.

Brazelton, T.B., E.R. Christophersen, A.C. Frauman, P.A. Gorski, J.M. Poole, A.C. Stadtler and C.L. Wright (1999), 'Instruction, timeliness, and medical influences affecting toilet training', *Pediatrics*, **103**, 1353–8.

Broderick, J. (2009), 'Voluntary carbon offsetting for air travel', in S. Gossling and P. Upham (eds), *Climate Change and Aviation: Issues, Challenges and Solutions*, London: Earthscan.

Brown, S. and G. Walker (2008), 'Understanding heat wave vulnerability in nursing and residential homes', *Building Research & Information*, **36**, 363–72.

Bryant, R.L. (2009), 'Peering into the abyss: environment, research and absurdity in the "age of stupid"' King's College London Department of

Geography, environment, politics and development working paper series, no. 19.

Bryson, B. (2010), *At Home: A Short History of Private Life*, London: Doubleday.

Buehlmann, U. and A. Schuler (2009), 'The U.S. household furniture industry: status and opportunities', *Forest Products Journal*, **59**.

Bulkeley, H. (2005), 'Reconfiguring environmental governance: towards a politics of scales and networks', *Political Geography*, **24**, 875–902.

Burgess J., T. Bedford, K. Hobson, G. Davies and C. Harrison (2003), '(Un)sustainable consumption', in I. Scoones and M. Leach (eds), *Negotiating Environmental Change*, Cheltenham, UK and Northampton, MA, USA: Edward Elgar, pp. 261–91.

Bushway, L.J., J.L. Dickinson, R.C. Stedman, L.P. Wagenet and D.A. Weinstein (2011), 'Benefits, motivations, and barriers related to environmental volunteerism for older adults: developing a research agenda', *International Journal of Aging and Human Development*, **72** (3), 189–206.

Cameron, J. and R. Gordon (2010), 'Building sustainable and ethical food futures through economic diversity: options for a mid-sized city', paper presented at the policy workshop on The Future of Australia's Mid-Sized Cities, Bendigo: Latrobe University, 28 and 29 September.

Canning, L. (2006), 'Rethinking market connections: mobile phone recovery, reuse and recycling in the UK', *Journal of Business & Industrial Marketing*, **21**, 320–29.

Canning, L. and I. Szmigin (2010), 'Death and disposal: the universal, environmental dilemma', *Journal of Marketing and Management*, **26**, 1129–42.

CanyonSnow Consulting (2009), 'Environmental benefits of thin computing', accessed 24 July 2012 at www.wyse.com/sites/default/files/documents/whitepapers/Wyse_Environmental_Benefits_WhitePaper.pdf.

Carrier, J. (1993), 'The rituals of Christmas giving', in Daniel Miller (ed.), *Unwrapping Christmas*, Oxford: Clarendon Press, pp. 55–74.

Carthey J., V. Chandra and M. Loosemore (2009), 'Adapting Australian health facilities to cope with climate-related extreme events', *Journal of Facilities Management*, **7**, 36–51.

Casado-Diaz, M.A. (2006), 'Retiring to Spain: an analysis of differences among north European nationals', *Journal of Ethnic and Migration Studies*, **32**, 1321–39.

Castree, N. (2004), 'The geographical lives of commodities: problems of analysis and critique', *Social & Cultural Geography*, **5**, 21–35.

Chappells, H. and E. Shove (2005), 'Debating the future of comfort: environmental sustainability, energy consumption and the indoor environment', *Building Research & Information*, **33**, 32–40.

Cheal, D. (1988), *The Gift Economy*, London: Routledge.

Cheng, J.C. and M.C. Monroe (2012), 'Connections to nature: children's affective attitude toward nature', *Environment and Behaviour*, **44**, 31–49.

Choice Australia (2009), 'Test HD set-top boxes – HD: highly disappointing', *Choice*, February, **59**.

Christophers, B. (2009), *Envisioning Media Power: On Capital and Geographies of Television*, Lanham, MD: Lexington.

Clarke, A (2000), '"Mother swapping": the trafficking of nearly new children's wear', in P. Jackson, M. Lowe, D. Miller and F. Mort (eds), *Commercial Cultures: Economies, Practices, Spaces*, Oxford: Berg, pp. 85–100.

Clarkson, J. (2004), 'Top Gear: Series 5, Episode 6', accessed 18 July 2012 at www.tv.com/shows/top-gear/series-5-episode-7-408901/.

Claudio, L. (2007), 'Waste couture: environmental impact of the clothing industry', *Environmental Health Perspective*, **115**, 449–54.

Clayden, A. and K. Dixon (2007), 'Woodland burial: memorial arboretum versus natural native woodland', *Mortality*, **12**, 240–60.

Clean Energy Council (2011), *Solar Hot Water and Heat Pump Study*, Melbourne, VIC: Clean Energy Council.

Climate Commission (Australia) (2012), 'The critical decade: international action on climate change', accessed 13 September 2012 at http://climate-commission.gov.au/wp-content/uploads/climatecommission_internationalReport_20120821.pdf.

Coen-Pirani, D., A. Leon and S. Lugauer (2008), 'The effect of household appliances on female labour force participation: evidence from micro data', Tepper School of Business, accessed 16 June 2012 at http://repository.cmu.edu/tepper/54.

Cohen, S.A. and J.E.S. Higham (2011), 'Eyes wide shut? UK consumer perceptions on aviation climate impacts and travel decisions to New Zealand', *Current Issues in Tourism*, **14**, 323–35.

Cohen, S.A., J.E.S. Higham and C.T. Cavaliere (2011), 'Binge flying. Behavioural addiction and climate change', *Annals of Tourism Research*, **38**, 1070–89.

Collier, K. (2012), 'Hot-water bottles are a must in cold times', *Daily Telegraph*, 16 August 2012.

Cook, I. (2006), 'Geographies of food: following', *Progress in Human Geography*, **30** (5), 655–66.

Copping, J. (2011), 'Crematoriums add corpse power to electricity grid', *Sydney Morning Herald*, 28 December.

Corbett, S. (2008), 'Can the cellphone help end global poverty?' *New York Times Magazine*.

Corbin, A. (1986), *The Foul and the Fragrant*, Leamington Spa: Berg.

Cordell, D., J-O. Drangert and S. White (2009), 'The story of phosphorus: global food security and food for thought', *Global Environmental Change*, **19**, 292–305.

Cordella, M., O. Wolf, A. Chapman and K. Bojczuk (2012), *Revision of the EU Ecolabel Criteria for Bed Mattresses*, European Commission, Institute for Prospective Technological Studies, Oakdene Hollins Research and Consulting.

Cowan, R.S. (1983), *More Work for Mother: The Ironies of Household Technology from the Open Hearth to the Microwave*, New York: Basic Books.

Cowan, R.S. (1985), 'How the refrigerator got its hum', in D. MacKenzie and J. Wajcman (eds), *The Social Shaping of Technology. How the Refrigerator Got its Hum*, Milton Keynes: Open University Press, pp. 202–218.

Craw, C. (2008), 'The ecology of emblem eating: environmentalism, nationalism and the framing of kangaroo consumption', *Media International Australia*, **127**, 82–94.

Cresswell, T. (2006), 'The right to mobility: the production of mobility in the courtroom', *Antipode*, **38**, 735–54.

CSIRO (Commonwealth Scientific and Industrial Research Organisation) (2011), *Flight Path to Sustainable Aviation: Towards Establishing a Sustainable Aviation Fuels Industry in Australia and New Zealand*, Australia: CSIRO.

Curtis, V. and A. Biran (2001), 'Dirt, disgust and disease', *Perspectives in Biology and Medicine*, **44**, 17–31.

Davidson, M. (2010), 'Sustainability as ideological praxis: the acting out of planning's master signifier', *City*, **14**, 390–405.

Davidson, M. (2012), 'Sustainable city as fantasy', *Human Geography*, **5**, 112–24.

Davies, A. (2011), 'On constructing ageing rural populations: "capturing" the grey nomad', *Journal of Rural Studies*, **27**, 191–9.

Davies, A. and A. James (2011), 'Environmental implications of an aging population', in A. Davies and A. James (eds), *Geographies of Ageing: Social Processes and the Spatial Unevenness of Population Ageing*, Farnham: Ashgate, pp. 159–73.

Davies, D.J. (2005), *A Brief History of Death*, Oxford: Blackwell.

Davies, J. (1999), 'Of caribou and kangaroo: agreements about wildlife', *Indigenous Law Bulletin*, **4** (21), 27–9.

Davis, L.W. (2008), 'Durable goods and residential demand for energy and water: evidence from a field trial', *RAND Journal of Economics*, **30** (2), 530–46.

Davison, A. (2011), 'A domestic twist on the eco-efficiency turn: environmentalism, technology, home', in R. Lane and A. Gorman-Murray (eds), *Material Geographies of Household Sustainability*, Farnham: Ashgate, pp. 35–50.

Davison, G. (2008), 'Down the gurgler: historical influences on Australian domestic water consumption', in P. Troy (ed.), *Troubled Waters: Confronting the Water Crisis in Australia's Cities*, Canberra: ANU E Press, pp. 37–66.

DCCEE (Department of Climate Change and Energy Efficiency) (2012a), 'Hot water heaters', Canberra: DCCEE, accessed 23 July 2012 at www.climatechange.gov.au/en/what-you-need-to-know/appliances-and-equipment/hot-water-systems.aspx.

DCCEE (2012b), 'Install a solar hot water system', Canberra: DCCEE, accessed 14 September 2012 at www.livinggreener.gov.au/energy/hot-water-systems/install-solar-hot-water.

Dent, B.B., S.L. Forbes and B.H. Stuart (2004), 'Review of human decomposition processes in soil', *Environmental Geology*, **45**, 576–85.

Department for Environment, Food and Rural Affairs (2011), 'Guidelines to Defra/DECC's Greenhouse Gas Conversion Factors for Company Reporting', United Kingdom Department for Environment, Food and Rural Affairs, August, accessed 15 May 2012 at www.defra.gov.uk/environment/economy/business-efficiency/reporting/.

Derrick, J.W. and D.C. Dumaresq (1999), 'Soil chemical properties under organic and conventional management in southern New South Wales', *Australian Journal of Soil Research*, **27**, 1047–55.

De Vet, E. (2012), *Australian Office Air Conditioning – Addiction or Worker Affliction?*, Conversations with AUSCCER accessed 21 December 2012 at http://uowblogs.com/ausccer/2012/10/12/australian-office-air-conditioning-addiction-or-worker-affliction/.

Devries, M.W. and M.R. Devries (1977), 'Cultural relativity of toilet training readiness: a perspective from East Africa', *Pediatrics*, **60**, 170–77.

DEWHA (Department of the Environment, Water, Heritage and the Arts) (2008), 'Energy use in the Australian residential sector 1986–2020', Canberra: DEWHA.

DFT (Department for Transport) (no date), 'Factsheets: UK transport greenhouse gas emissions', accessed 15 May 2011 at www.dft.gov.uk/statistics/series/energy-and-environment/.

Dickinson, J.E., L.M. Lumsdon and D. Robbins (2011), 'Slow travel: issues for tourism and climate change', *Journal of Sustainable Tourism*, **19**, 281–300.

Diesendorf, M. (2007), *Paths to a Low-Carbon Future*, Epping, NSW: Sustainability Centre.

DITRDLG (Department of Infrastructure, Transport, Regional Development and Local Government) (2012), 'Top sellers & performers online: March 2012', Canberra: DITRDLG, accessed 6 August 2012 at www.greenvehicleguide.gov.au.

Dodge, M. and R. Kitchin (2012), 'Towards touch-free spaces: sensors, software and the automatic production of shared public toilets', in M. Paterson and M. Dodge (eds), *Touching Space, Placing Touch*, Farnham: Ashgate.

Doeringer, P. and S. Crean (2006), 'Can fast fashion save the US apparel industry?', *Socio–Economic Review*, **4**, 353–77.

Dombroski, K (2012), 'Poor mothers are not poor mothers: cross-cultural learning between northwest China and Australasia', accessed 28 May 2012 at www.communityeconomies.org/site/assets/media/Kelly_Dombroski/DOMBROSKI2C20CCR20POSTGRAD20CONFERENCE1.pdf.

Donaldson, B., B. Nagengast and G. Meckler (1994), *Heat and Cold. Mastering the Great Indoors: A Selective History of Heating, Ventilation and Air Conditioning*, Atlanta, GA: American Society of Heating, Refrigeration and Air-Conditioning Engineers (ASHRAE) Transactions.

DOT (US Department of Transportation) (2007), 'Air travel for business vs other purposes', Washington, DC:, accessed 26 September 2012 at https://ntl.custhelp.com/app/answers/detail/a_id/252/~/percentage-of-air-travel-for-business-vs-other-purposes.

Doteuchi, A. (2008). *'Downsizing' of Housing and Lifestyles for a Low-Carbon Aging Society*, Social Development Research Group, NLI Research Institute.

Douglas, M. (1984 [1966]), *Purity and Danger. An Analysis of the Concepts of Pollution and Taboo*, London: Ark.

Dowling, R. (2000), 'Cultures of mothering and car use in suburban Sydney: a preliminary investigation', *Geoforum*, **31**, 345–53.

Dowling, R. (2010), 'Geographies of identity: climate change, governmentality and activism', *Progress in Human Geography*, **34**, 488–95.

Dowling, R. and E. Power (2011), 'Beyond McMansions and green homes: thinking household sustainability through materialities of homeyness', in R. Lane and A. Gorman-Murray (eds), *Material Geographies of Household Sustainability*, Aldershot: Ashgate.

Druckman, A. and T. Jackson (2009), 'The carbon footprint of UK households 1990–2004: a socio-economically disaggregated, quasi-multi-regional input-output model', *Ecological Economics*, **68**, 2066–77.

DSEWPC (Department of Sustainability, Environment, Water, Population and Communities) (2012), 'Plastic bags', Canberra: DSEWPC, accessed 7 July 2012 at www.environment.gov.au/settlements/waste/plastic-bags/index. html.

EC (European Commission) (2008), *Furniture: Background Product Report*, European Commission Green Public Procurement Training Toolkit, Barcelona: ICLEI and Ecoinstitut.

Edensor, T. (2004), 'Automobility and national identity: representation, geography and driving practice', *Theory, Culture & Society*, **21**, 101–20.

Edensor, T. (ed.) (2010), *Geographies of Rhythm: Nature, Place, Mobilities and Bodies*, Farnham: Ashgate.

Edensor, T. and S. Millington (2009), 'Illuminations, class identities and the contested landscapes of Christmas', *Sociology*, **43**, 103–21.

Edwards, C. and J. Fry (2011), *Life Cycle Assessment of Supermarket Carrier Bags: A Review of the Bags Available in 2006*, report SC030148, Bristol: Environment Agency.

EIA (Energy Information Administration) (2011), 'Air conditioning in nearly 100 million U.S. homes', press release, 19 August, Washington: EIA, accessed 27 September 2012 at www.eia.gov/consumption/residential/reports/2009/air-conditioning.cfm.

Elliot, R., S. Eccles and M. Ritson (1996), 'Reframing IKEA: commodity signs, consumer creativity and the social/self dialectic', *Advances in Consumer Research*, **23**, 127–31.

Energy Saving Trust (2011a), *A Buyer's Guide to Solar Water Heating*, London: Energy Saving Trust.

Energy Saving Trust (2011b), *Here Comes the Sun: A Field Trial of Solar Water Heating Systems*, London: Energy Saving Trust.

Environmental Leader (2007), 'Lexus hybrid ad banned for misleading environmental claims', 24 May, accessed 17 August 2012 at www.environmentalleader.com/2007/05/24/lexus-hybrid-ad-banned-for-misleading-environmental-claims/.

EPA (Environmental Protection Agency) (2006), *Protecting Water Resources with High Density Developments*, Washington, DC: EPA.

EPA (2008), 'Fact sheet: management of electronic waste in the United States', Washington, DC: EPA, accessed 14 September 2012 at www.epa.gov/epawaste/conserve/materials/ecycling/docs/fact7–08.pdf.

EPA (2012a), 'Green vehicle guide. You have green options!' Washington, DC: EPA, accessed 13 May 2012 at http://nepis.epa.gov.

EPA (2012b), 'eCycle cell phones', Washington, DC: EPA, accessed 23 May 2012 at www.epa.gov/epawaste/partnerships/plugin/cellphone/index.htm.

ERC (Environment and Resources Committee) (2011), *The Environmental Impacts of Conventional Burials and Cremations*, Issues Paper No. 3, Queensland Parliament: ERC, accessed 31 July 2012 at www.parliament.qld.gov.au/Documents/.

Eternal Reefs (no date), 'Eternal reefs', accessed 15 November 2011 at www.eternalreefs.com/index.html.

EuroHeat (2008), *Improving Public Health Responses to Extreme Weather/Heat Waves – EuroHeat*, Copenhagen: World Health Organization.

Everts, S. (2010), 'Green for eternity start-up companies introduce two routes to stay environmentally friendly after you're dead and gone', *Chemical and Engineering News*, **88** (26), 41–2.

Falkenmark, M. (2008), 'Water and sustainability: a reappraisal', *Environment*, **50** (2), 5–16.

Farbotko, C and G. Waitt (2011), 'Residential air-conditioning and climate change: voices of the vulnerable', *Health Promotion Journal of Australia*, **22**, 13–16.

Fargione, J., J. Hill, D. Tilman, S. Polasky and P. Hawthorne (2008), 'Land clearing and the biofuel carbon debt', *Science*, **319** (5867), 1235–8.

Featherstone, M., N. Thrift and J. Urry (eds) (2004), *Automobilities*, London: Sage.

Fehske, A., G. Fettweis, J. Malmodin and G. Biczok (2011), 'The global footprint of mobile communications: the ecological and economic perspective', *Communications Magazine*, **49** (8), 55–62.

Feng, A., D. Gordon, H. Hui, D. Kodjak and D. Rutherford (2007), 'Passenger vehicle greenhouse gas and fuel economy standards: a global update', The International Council on Clean Transportation, accessed 17 June 2012 at www.theicct.org/passenger-vehicle-greenhouse-gas-and-fuel-economy-standards.

Fiedler, S., J. Breuer, C.M. Pusch, S. Holley, J. Wahl, J. Ingwersen and M. Graw (2012), 'Graveyards – special landfills', *Science of The Total Environment*, **419**, 90–97.

Fine, B. and E. Leopold (1993), *The World of Consumption*, London and New York: Routledge.

FIRA (Furniture Industry Research Association) (2011), 'Furniture carbon footprinting', Stevenage: FIRA, accessed 17 September 2012 at www.fira.co.uk/document/fira-carbon-footprinting-document-2011.pdf.

Fishbein, B.K. (2002), 'Waste in the wireless world: the challenge of cell phones', accessed 1 June 2012 at www.informinc.org/reportpdfs/wp/WasteintheWirelessWorld.pdf.

Fletcher, K. (2008), *Sustainable Fashion and Textiles*, London: Earthscan.

Folger, T. (2011), 'The secret ingredients of everything', *National Geographic*, **219** (6), 136–45.

Forrest, N. and E. Williams (2010), 'Life cycle environmental implications of residential swimming pools', *Environment, Science and Technology*, **44**, 5601–7.

Foster, J. (1993), 'Solar water heating in Queensland: the roles of innovation attributes, attitudes and information in the adoption process', *Prometheus*, **11**, 219–33.

Frapart, S. (2010), 'Cozy winter fires – carbon impact', *Green Blizzard*, 24 November, accessed 11 July 2012 at http://greenblizzard.com/carbon-footprint/2010/11/24/fireplaces/.

Freeman, J. (2004), *The Making of the Modern Kitchen; Gender, Design and Culture in the Twentieth Century*, Oxford: Berg.

Frey, S.D., D.J. Harrison and E.H. Billet (2006), 'Ecological footprint analysis applied to mobile phones', *Journal of Industrial Ecology*, **10**, 199–216.

Frumkin, H., L. Frank and R. Jackson (2004), *Urban Sprawl and Public Health*, Washington, DC: Island Press.

Fu, X., T.H. Oum and A. Zhang (2010), 'Air transport liberalization and its impacts on airline competition and air passenger traffic', *Transportation Journal*, **49** (4), 24–41.

Gandy, M. (2006), 'The bacteriological city and its discontents', *Historical Geography*, **34**, 14–25.

Gardiner, A. (2010), 'Do rainwater tanks herald a cultural change in household water use?', *Australasian Journal of Environmental Management*, **17**, 100–111.

Garnaut, R. (2008), *Climate Change Review*, Canberra: Australian Government Publishing Service.

Garnett, T. (2011), 'Where are the best opportunities for reducing greenhouse gas emissions in the food system (including the food chain)?', *Food Policy*, **36**, S23–S32.

Gaston, K.J., P.H. Warren, K. Thompson and R.M. Smith (2005), 'Urban domestic gardens (IV): the extent of the resource and associated features', *Biodiversity and Conservation*, **14**, 3327–49.

Gaynor, A. (2001), '*Harvest of the suburbs: an environmental history of suburban food production in Perth and Melbourne', 1880–2000*, PhD thesis, University of Western Australia.

Gazzard, J. (2008), 'Bio-fuelled or bio-fooled?', *Aviation and the Environment*, 44–9.

George Wilkenfeld and Associates (2010), *Regulation Impact Statement: for Decision Phasing Out Greenhouse-Intensive Water Heaters in Australian Homes*, Sydney, NSW: National Framework for Energy Efficiency.

Geppert, J. and R. Stamminger (2010), 'Do consumers act in a sustainable way using their refrigerator? The influence of consumer real life behaviour on the energy consumption of cooling appliances', *International Journal of Consumer Studies*, **34**, 219–27.

Ger, G. and R.W. Belk (1999), 'Accounting for materialism in four cultures', *Journal of Material Culture*, **4**, 183–204.

Gerbens-Leenes, P.W., H.C. Moll and A.J.M. Schoot Uiterkamp (2003), 'Design and development of a measuring method for environmental sustainability in food production systems', *Ecological Economics*, **46**, 231–48.

Geyer, R. and V. Blass (2010), 'The economics of cell phone reuse and recycling', *The International Journal of Advanced Manufacturing Technology*, **47**, 515–25.

Ghosh, S. and L. Head (2009), 'Retrofitting the suburban garden: morphologies and some elements of sustainability potential of two Australian residential suburbs compared', *Australian Geographer*, **40**, 319–46.

Gibson, C. (2012), 'Wood fires in the suburbs: affordability, a retro trip, or reconnecting with nature?', conversations with AUSCCER, 12 July, accessed 18 September 2012 at http://uowblogs.com/ausccer/2012/07/12/wood-fires-in-the-suburbs-affordability-a-retro-trip-or-reconnecting-with-nature/.

Gibson, C. and Stanes, E. (2010), 'Is green the new black? Exploring ethical fashion consumption', in T. Lewis and E. Potter (eds), *Ethical Consumption: A Critical Introduction*, London and New York: Routledge, pp. 169–85.

Gibson, C., L. Head, N. Gill and G. Waitt (2010), 'Climate change and household dynamics: beyond consumption, unbounding sustainability', *Transactions of the Institute of British Geographers*, **36**, 3–8.

Gibson, C., G. Waitt, L. Head and N. Gill (2011), 'Is it easy being green? On the dilemmas of material cultures of household sustainability', in R. Lane and A. Gorman-Murray (eds), *Material Geographies of Household Sustainability*, Farnham: Ashgate, pp. 19–33.

Gille, Z. and S. O'Riain (2002), 'Global ethnography', *Annual Review of Sociology*, **28**, 271–95.

Glance, D (2011), 'The Internet of things: this is where we're going', accessed 27 September 2012 at http://theconversation.edu.au/the-internet-of-things-this-is-where-were-going-3965.

Gleeson, B. (2006), *Australian Heartlands: Making Space for Hope in Suburbs*, Sydney, NSW: Allen & Unwin.

Gleeson, B. (2008), 'Waking from the dream: an Australian perspective on urban resilience', *Urban Studies*, **45**, 2653–68.

Gleick, P. (2010), 'Roadmap for sustainable water sources in southwestern North America', *Proceedings of the National Academy of Sciences*, **107**, 21300–305.

Glew, D., L.C. Stringer, A.A. Acquaye and S. McQueen-Mason (2012), 'How do end of life scenarios influence the environmental impact of product supply chains? Comparing biomaterial and petrochemical products', *Journal of Cleaner Production*, **29–30**, 122–31.

Gobillon, L. and F-C. Wolff (2011), 'Housing and location choices of retiring households: evidence from France', *Urban Studies*, **48**, 331–47.

GoodGuide (2012), 'Best cell phones ratings', accessed 12 June 2012 at www.goodguide.com/categories/332304-cell-phones##products.

Gossling, S. and P. Upham (eds) (2009), *Climate Change and Aviation: Issues, Challenges and Solutions*, London: Earthscan, pp. 131–51.

Gossling, S., J-P. Ceron, G. Dubois and C.M. Hall (2009a), 'Hypermobile travellers', in S. Gossling, S., L. Haglund, H. Kallgren, M. Revahl and J. Hultman (2009b), 'Swedish air travellers and voluntary carbon offsets: towards the co-creation of environmental value?', *Current Issues in Tourism*, **12** (1), 1–19.

Gossling, S., J. Broderick, P. Upham, J. Ceron, G. Dubois, P. Peeters and W. Strasdas (2007), 'Voluntary carbon offsetting schemes for aviation: efficiency, credibility and sustainable tourism', *Journal of Sustainable Tourism*, **15**, 223–48.

Gregg, M. (2011), *Work's Intimacy*, London: Polity.

Gregson, N. (2007), *Living With Things: Ridding, Accommodation, Dwelling*, Wantage: Sean Kingston.

Gregson, N., A. Metcalfe and L. Crewe (2007), 'Identity, mobility, and the throwaway society', *Environment and Planning D*, **25**, 682–700.

Grieve, C., R. Lawson and J. Henry (2012), 'Understanding the non-adoption of energy efficient hot water systems in New Zealand', *Energy Policy*, **48**, 369–73.

Grigg, G. (ed.) (1995), *Conservation Through Sustainable Use of Wildlife*, University of Queensland: Centre for Conservation Biology.

Grigg, G.C. (1987), 'Kangaroo harvesting: a new approach', *Australian Natural History*, **22**, 204–5.

Gudmundsson, S.V. and A. Anger (2012), 'Global carbon dioxide emissions scenarios for aviation derived from IPCC storylines: a meta-analysis', *Transportation Research D*, **17**, 61–5.

Gustafson, P. (2001), 'Retirement migration and transnational lifestyles', *Ageing & Society*, **21**, 371–94.

Gustafson, P. (2008), 'Transnationalism in retirement migration: the case of North European retirees in Spain', *Ethnic and Racial Studies*, **31**, 451–75.

Guthman, J. (2003), 'Fast food/organic food: reflexive tastes and the making of "yuppie chow"', *Social & Cultural Geography*, **4**, 45–58.

Hackett, M.J. and N.F. Gray (2009), 'Carbon dioxide emission savings potential of household water use reduction in the UK', *Journal of Sustainable Development*, **2**, 36–43.

Hakansson, H. and A. Waluszewski (2002), *Managing Technological Development: IKEA, the Environment and Technology*, London: Routledge.

Hall, C.A.S, R.J. Pontius, L. Coleman and J-Y. Ko (1994), 'The environmental consequences of having a baby in the United States', *Population and Environment*, **15**, 505–24.

Hamilton, C., R. Dennis and D. Baker (2005), *Wasteful Consumption in Australia*, Canberra: Australia Institute.

Hammad, M.A. and M.A. Alsaad (1999), 'The use of hydrocarbon mixtures as refrigerants in domestic refrigerators', *Applied Thermal Engineering*, **19**, 1181–9.

Handfield, R.B., S.V. Walton, L.K. Seegers and S.A. Melnyk (1997), '"Green" value chain practices in the furniture industry', *Journal of Operations Management*, **15**, 293–315.

Haq, G., A. Owen, E. Dawkins and J. Barrett (2007), *The Carbon Cost of Christmas*, Stockholm: Stockholm Environment Institute.

Harrington, L. and M. Damnics (2004), *Energy Labelling and Standards Programs Throughout the World*, National Appliance and Equipment Energy Efficiency Committee, Australia, accessed 13 September 2012 at www.energyrating.com.au/library/pubs/200404-internatlabelreview.pdf.

Harris, M. (2007), *Grave Matters: A Journey Through the Modern Funeral Industry to a Natural Way of Burial*, New York: Scribner.

Harvey, M. and S. Pilgrim (2011), 'The new competition for land: food, energy, and climate change', *Food Policy*, **36**, S40–S51.

Hasse, D. and H. Nuissl (2007), 'Does urban sprawl drive changes in water balance and policy? The case of Leipzig (Germany) 1870–2003', *Landscape and Urban Planning*, **80**, 1–13.

Hawkins, G. (2006), *The Ethics of Waste: How We Relate to Rubbish*, Oxford: Rowman and Littlefield.

Hayes-Conroy, A. and J. Hayes-Conroy (2008), 'Taking back taste: feminism, food and visceral politics', *Gender, Place & Culture*, **15**, 461–73.

Head, L. (2012), 'Clothing as adaptive strategy', conversations with AUSC-CER, 1 August 2012, accessed 18 September 2012 at http://uowblogs.com/ausccer/2012/08/01/clothing-as-adaptive-strategy/.

Head, L. and P. Muir (2007a), *Backyard. Nature and Culture in Suburban Australia*, Wollongong, NSW: University of Wollongong Press.

Head, L. and P. Muir (2007b), 'Changing cultures of water in eastern Australian backyard gardens', *Social and Cultural Geography*, **8**, 889–906.

Head, L., J. Atchison and A. Gates (2012), *Ingrained. A Human Bio-geography of Wheat*, Farnham: Ashgate.

Head, L., P. Muir and E. Hampel (2004), 'Australian backyard gardens and the journey of migration', *The Geographical Review*, **94**, 326–47.

Healy, S. (2008), 'Air-conditioning and the "homogenization" of people and built environments', *Building Research & Information*, **36**, 312–22.

Hetherington, K. (2004), 'Secondhandness: consumption, disposal and absent presence', *Environment and Planning D: Society and Space*, **22**,157–73.

Hill, A. (2011), 'A helping hand and many green thumbs: local government, citizens and the growth of a community-based food economy', *Local Environment*, **16**, 539–53.

Hinchliffe, S. (1996), 'Helping the earth begins at home. The social construction of socio-environmental responsibilities', *Global Environmental Change*, **6**, 53–62.

Hinchliffe, S. (1997), 'Locating risk: energy use, the "ideal" home and the non-ideal world', *Transactions of the Institute of British Geographers*, **22**, 197–209.

Hischier, R and I. Baudin (2010), 'LCA study of a plasma television device', *International Journal of Life Cycle Assessment*, **15** (5), 428–38.

Hitchings, R. (2010), 'Seasonal climate change and the indoor city worker', *Transactions of the Institute of British Geographers*, **35**, 282–98.

Hitchings, R. and S.J. Lee (2008), 'Air conditioning and the material culture of routine human encasement: the case of young people in contemporary Singapore', *Journal of Material Culture*, **13**, 251–65.

Hobson, K. (2002), 'Competing discourses of sustainable consumption: does the "rationalisation of lifestyles" make sense?', *Environmental Politics*, **11**, 95–120.

Hobson, K. (2003), 'Thinking habits into action: the role of knowledge and process in questioning household consumption practices', *Local Environment*, **8**, 95–112.

Hobson, K. (2006), 'Bins, bulbs, and shower timers: on the "techno-ethics" of sustainable living', *Ethics, Place and Environment*, **9**, 317–36.

Hobson, K. (2008), 'Reasons to be cheerful: thinking sustainably in a (climate) changing world', *Geography Compass*, **2**, 199–214.

HoC (House of Commons) (2002), *Disposal of Refrigerators Report*, proceedings of the Committee, minutes of evidence and appendices, fourth report of session 2001–2, 20 June, London: House of Commons Environment Food and Rural Affairs Committee.

Hoekstra, A.Y. and A.K. Chapagain (2007), 'Water footprints of nations: water use by people as a function of their consumption pattern', *Water Resources Management*, **21**, 35–48.

Hogarth, M. (2008), *Crunch Time for Offsets: Where to Now for the Voluntary Carbon Market?*, Sydney, NSW: T.E. Centre.

Howarth, G. (2007), *Death and Dying: A Sociological Introduction*, Cambridge: Polity.

Howe, C., R.N. Jones, S. Maheepala and B. Rhodes (2005), 'Implications of potential climate change for Melbourne's water resources, a collaborative project between Melbourne Water and CSIRO Urban Water and Climate Impact Groups', accessed 23 January 2012 at www.melbournewater. com.au/content/sustainability/climate_change/climate_change.asp.

Hu, F.B., T.Y. Li, G.A. Colditz, W.C. Willett and J.E. Manson (2003), 'Television watching and other sedentary behaviors in relation to risk of obesity and type 2 diabetes mellitus in women', *Journal of the American Medical Association*, **289**, 1785–91.

Huang, E.M. and K.N. Truong (2008), *Breaking the Disposable Technology Paradigm: Opportunities for Sustainable Interaction Design for Mobile Phones*, the 26th annual SIGCHI conference on human factors in computing systems Florence, Italy, accessed 24 June 2012 at www.chi2008.org/.

Huang, E.M., K. Yatani, K.N. Truong, K. A. Kientz and S.N. Patel (2009), *Understanding Mobile Phone Situated Sustainability: The Influence of Local Constraints and Practices on Transferability*, Pervasive Computing, IEEE, pp. 46–53.

Hughner, R.S., P. McDonagh, A. Prothero, C.J. Shultz and J. Stanton (2007), 'Who are organic food consumers? A compilation and review of why people purchase organic food', *Journal of Consumer Behaviour*, **6**, 94–110.

Hulme, M. (2007), 'Newspaper scare headlines can be counter-productive', *Nature*, **445**, 818.

Hulme, M. (2008), 'Geographical work at the boundaries of climate change', *Transactions of the Institute of British Geographers*, **33**, 5–11.

Hulme, M. (2010), 'Cosmopolitan climates: hybridity, foresight and meaning', *Theory Culture Society*, **27**, 267–76.

Hungerford, D. (2004), *Living Without Air-Conditioning in a Hot Climate: Thermal Comfort in Social Context*, American Council for an Energy Efficient Economy Summer Study on Energy Efficiency in Building Proceedings, accessed 27 August 2012 at http://aceee.org/proceedings-paper/ss04/panel07/paper11.

Hustvedt G. (2011), 'Review of laundry energy efficiency studies conducted by the US Department of Energy', *International Journal of Consumer Studies*, **35**, 228–36.

Hyder Consulting (2008), *Plastic Retail Carry Bag Use 2006 and 2007 Consumption*, Environment Protection and Heritage Council, Melbourne, VIC: Hyder Consulting.

IATA (International Air Transport Association) (2011), *2011 Report on Alternative Fuels*, Montreal: IATA.

IBISWorld (2012), 'Furniture retailing market research report', accessed 14 September 2012 at www.ibisworld.com.au/industry/default.aspx? indid= 411.

ICAO (International Civil Aviation Organisation) (2010), *Carbon Emissions Calculator 2010: Version Three*, Montreal: ICAO.

IEA (International Energy Agency) (2003), *Cool Appliances. Policy Strategies for Energy-Efficient Homes*, Paris: IEA.

IEA (2009a), 'Transport, energy and CO_2: moving towards sustainability', Paris: IEA, accessed 15 June 2012 at www.iea.org/Textbase/npsum/transport2009SUM.pdf.

IEA (2009b), 'IEA expects energy use by new electronic devices to triple by 2030 but sees considerable room for more efficiency', Paris: IEA, accessed 29 May 2012 at www.iea.org/press/pressdetail.asp?PRESS_REL_ID=284

Isenstadt, S. (1998), 'Visions of plenty: refrigerators in America around 1950', *Journal of Design History*, **11**, 311–21.

ITF (International Transport Forum) (2009), *The Cost and Efficiency of Reducing Transport Greenhouse Gas Emissions*, Preliminary Findings, Paris: ITF, OECD.

ITU (International Telecommunications Union) (2011), 'The world in 2011: ICT facts and figures', Geneva: ITU, accessed 31 May 2012 at www.itu.int/ITU-D/ict/facts/2011/material/ICTFactsFigures2011.pdf.

Jalland, P. (2006), *Changing Ways of Death in Twentieth-Century Australia*, Sydney, NSW: University of New South Wales Press Ltd.

Jarvis, H. (2011), 'Saving space, sharing time: integrated infrastructures of daily life in cohousing', *Environment and Planning A*, **43**, 560–77.

Jewitt, S. (2011), 'Geographies of shit: spatial and temporal variations in attitudes towards human waste', *Progress in Human Geography*, **35** (5), 608–26.

Johnson, E. (1998), 'Global warming from HFC', *Environmental Impact Assessment Review*, **18**, 485–92.

Johnson, L. and J. Lloyd (2004), *Sentenced to Everyday Life – Feminism and the Housewife*, Oxford: Berg.

Kahhat, R., J. Kim, M. Xu, B. Allenby, E. Williams and P. Zhang (2008), 'Exploring e-waste management systems in the United States', *Resources, Conservation and Recycling*, **52**, 955–64.

Kaika, M. (2004), 'Interrogating the geographies of the familiar: domesticating nature and constructing the autonomy of the modern home', *International Journal of Urban and Regional Research*, **28**, 265–86.

Kaika, M. (2005), *City of Flows. Modernity, Nature and the City*, London: Routledge.

Katayama, M. and R. Sugihara (2011), 'Which type of washing machine should you choose', *International Journal of Consumer Studies*, **35**, 237–42.

Katz, J. (ed.) (2011), *Mobile Communication: Dimensions of Social Policy*, New Brunswick, NJ: Transaction.

Katz, J. and S. Sugiyama (2006), 'Mobile phones as fashion statements: evidence from student surveys in the US and Japan', *New Media & Society*, **8**, 321–37.

Kaufmann, J.C. (1998), *Dirty Linen: Couples and Their Laundry*, London: Middlesex University Press.

Keijzer, E.E. (2011), *Environmental Impact of Funerals: Life Cycle Assessments of Activities After Life*, Masters Thesis, Utrecht, Netherlands: University of Groningen.

Keijzer, E.E. and H.J.G. Kok (2011), *Environmental Impact of Different Funeral Technologies*, TNO report-060-UT-2011-001432, Utrecht, Netherlands: TNO.

Keilman, N. (2003), 'The threat of small households', *Nature*, **421** (6922), 489–90.

Kellaher, L., D. Prendergast and J. Hockey (2005), 'In the shadow of the traditional grave', *Mortality*, **10**, 237–50.

Kelly, T. (2004), 'Water and sustainability', *Australian Planner*, **41**, 37–8.

Kenway, S.J., A. Priestley, S. Cook, S. Seo, M. Inman, A. Gregory and A. Hall (2008), *Energy Use in the Provision and Consumption of Urban Water in Australia and New Zealand*, Australia: CSIRO.

Kline, R. (2000), *Consumers in the Country: Technology and Social Change in Rural America*, Baltimore, MD: Johns Hopkins University Press.

Kline, R.R (2003), 'Home ideologies', in N.E Lerman, R. Oldenziel and A.P Mohun, *Gender and Technology*, Baltimore, MD: Johns Hopkins University Press, pp. 392–424.

Klinenberg, E. (1999), 'Denaturalizing disaster: a social autopsy of the 1995 Chicago heat wave', *Theory and Society*, **28**, 239–95.

Klinenberg, E. (2002), *Heat Wave: A Social Autopsy of Disaster in Chicago*, Chicago, IL: University of Chicago Press.

Klocker, N., C. Gibson and E. Borger (2012), 'Living together, but apart: material geographies of everyday sustainability in extended family households', *Environment and Planning A*, **44**, 2240–59.

Klocker, N. and L. Head (2013), 'Diversifying ethnicity in Australia's population and environment debates', *Australian Geographer*, **44**, 41–62.

Knowaste (2012), 'Nappy days – Scotland's nappies to be recycled with Knowaste', accessed 28 August 2012 at www.knowaste.com/news-and-events/nappy-days-scotland.

Knox-Hayes, J. (2010), 'Constructing carbon market spacetime: climate change and the onset of neo-modernity', *Annals of the Association of American Geographers*, **100**, 953–62.

Kollmuss, A. and J. Lane (2008), *Carbon Offsetting & Air Travel: Part 1: CO_2-Emissions Calculations*, Stockholm Environment Institute discussion paper, accessed 23 July 2012 at http://sei-us.org/publications/id/94.

Kollmuss, A., M. Lazarus, C. Lee, M. LeFranc and C. Polycarp (2010), *Handbook of Carbon Offset Programs: Trading Systems, Funds, Protocols and Standards*, London: Earthscan.

Kong, L. (2000), 'Nature's pleasures, nature's dangers. Urban children and the natural world', in S.L. Holloway and G. Valentine (eds), *Children's Geographies: Playing, Living, Learning*, Abingdon: Routledge, pp. 222–32.

Kovats, R.S. and S. Hajat (2008), 'Heat stress and public health: a critical review' *Annual Review of Public Health*, **29**, 41–55.

Kozak, G.L. and G.A. Keoleian (2003), *Printed Scholarly Books and E-Book Reading Devices: A Comparative Life Cycle Assessment of Two Book Options*, IEEE International Symposium on Electronics and the Environment, 19–22 May, DOI:10.1109/ISEE.2003.1208092.

LaForce, T. (2010), 'Love miles and carbon credits', accessed 21 May 2012 at http://blogs.nature.com/artesia/2010/12/22/love-miles-and-carbon-credits.

Laitala, K., C. Boks and I.G. Klepp (2011), 'Potential for environmental improvements in laundering', *International Journal of Consumer Studies*, **35**, 254–64.

Laitala, K., I.G. Klepp and C. Boks (2012), 'Changing laundry habits in Norway', *International Journal of Consumer Studies*, **36**, 228–237.

Lakoff, G. (2004), *Don't Think of an Elephant! Know Your Values and Frame the Debate*, White River Junction, VT: Chelsea Green.

Lally, E. (2002), *At Home with Computers*, Oxford: Berg.

Lane, R. and A. Gorman-Murray (eds) (2011), *Material Geographies of Household Sustainability*, Aldershot: Ashgate.

Lane, R. and M. Watson (2012), 'Stewardship of things: the radical potential of product stewardship for re-framing responsibilities and relationships to products and materials', *Geoforum*, accessed 6 September 2012 at http://dx.doi.org/10.1016/j.geoforum.2012.03.012.

Lane, R., R. Horne and J. Bicknell (2009), 'Routes of reuse of second-hand goods in Melbourne households', *Australian Geographer*, **40**, 151–68.

Lanvin, M. (1990), 'TV design' in M. Geller (ed.), *From Receiver to Remote Control: The TV Set*, New York: New Museum of Contemporary Art.

Larkins, R. (2007), *Funeral Rights: What the Australian "Death Care" Industry Doesn't Want You to Know*, Melbourne, VIC: Viking.

Lassen, C. (2010), 'Environmentalist in business class: an analysis of air travel and environmental attitude', *Transport Reviews*, **30**, 733–51.

Lassen, C., B.T. Laugen and P. Naess (2006), 'Virtual mobility and organisational reality – a note on the mobility needs in knowledge organisations', *Transportation Research Part D*, **11**, 459–63.

Latour, B. (1988), *The Pasteurization of France*, Cambridge: Harvard University Press.

Lawrence, K. and P. McManus (2008), 'Towards household sustainability in Sydney? Impacts of two sustainable lifestyle workshop programs on water consumption in existing homes', *Geographical Research*, **46**, 314–32.

Leichenko, R.M., K.L. O'Brien and W.D. Solecki (2010), 'Climate change and the global financial crisis: a case of double exposure', *Annals of the Association of American Geographers*, **100**, 963–72.

Leinwand, P., L.H. Moeller and K.B. Shriram (2008), *Consumer Spending in the Economic Downturn: The Wide Ranging Impact on Consumer Behaviour*, Chicago, IL: Booz & Company.

Lenzen, M., D. Moran, K. Kanemoto, B. Foran, L. Lobefaro and A. Geschke (2012), 'International trade drives biodiversity threats in developing nations', *Nature*, **486**, 109–12.

Lepawsky, J. (2012), 'Legal geographies of e-waste legislation in Canada and the US: jurisdiction, responsibility and the taboo of production', *Geoforum*, http://dx.doi.org/10.1016/j.geoforum.2012.03.006.

Leslie, D. and S. Reimer (2003), 'Fashioning furniture: restructuring the furniture commodity chain', *Area*, **35**, 427–37.

Leslie, D. and S. Reimer (2006), 'Situating design in the Canadian household furniture industry', *Canadian Geographer*, **50**, 319–41.

Levine, E. (1997), 'Jewish views and customs on death', in C.M. Parkes, P. Laungai and W. Young (eds), *Death and Bereavement Across Cultures*, London: Routledge, pp. 98–130.

Levi-Strauss, C. (1993), 'Father Christmas executed', in D. Miller (ed.), *Unwrapping Christmas*, Oxford: Clarendon, pp. 38–54.

Lewis, H., K. Verghese and L. Fitzpatrick (2010), 'Evaluating the sustainability impacts of packaging: the plastic carry bag dilemma', *Packaging Technology and Science*, **23**, 145–60.

Li, V.S. and S. Rosenthal (no date), 'Content and emission characteristics of artificial wax firelogs', accessed July 2012 at http://greenblizzard.com/wp-content/uploads/2010/11/Firelogs.pdf.

Li, X.D., C.S. Poon, S.C. Lee, S.S. Chung and F. Luk (2003), 'Waste reduction and recycling strategies for the in-flight services in the airline industry', *Resources, Conservation and Recycling*, **37** (2), 87–99.

LifeArt Australia (no date), 'LifeArt', accessed 8 December 2011 at www.lifeart.com.au/.

Lines-Kelly, R. (2010), *Soils ARE Dirt*, 19th World Congress of Soil Science, Soil Solutions for a Changing World, 1–6 August, Brisbane, QLD, Australia.

Ling, R. and N. Bashir (2011), 'Mobile communication and the environment', in J. Katz (ed.), *Mobile Communication: Dimensions of Social Policy*, New Brunswick, NJ: Transaction: pp. 75–86.

Liverman, D. (2008), 'Assessing impacts, adaptation and vulnerability: reflections on the Working Group II Report of the Intergovernmental Panel on Climate Change', *Global Environmental Change*, **18**, 4–7.

Loram, A., P.H. Warren and K.J. Gaston (2008), 'Urban domestic gardens (XIV): the characteristics of gardens in five cities', *Environmental Management*, **42**, 361–76.

Lorenzoni, I., S. Nicholson-Cole and L. Whitmarsh (2007), 'Barriers perceived to engaging with climate change among the UK public and their policy implications', *Global Environmental Change*, **17**, 445–59.

Low, K.E.Y. (2006), 'Presenting the self, the social body, and the olfactory: managing smells in everyday life experiences', *Sociological Perspective*, **49**, 607–31.

Luber, G. and N. Prudent (2009), 'Climate change and human health', *Transactions of the American Clinical and Climatological Association*, **120**, 113–17.

Luhrmann, M. (2009), 'Consumer expenditures and home production at retirement: new evidence from Germany', *German Economic Review*, **11**, 225–45.

Lundberg, S., R. Startz and S. Stillman (2003), 'The retirement-consumption puzzle: a marital bargaining approach', *Journal of Public Economics*, **87**, 1199–218.

Luntz, S. (2007), 'Earth to earth, not ashes to ashes', *Australasian Science*, **28** (5), 8.

Macdonald, G.M. (2010), 'Water, climate change, and sustainability in the southwest', *Proceedings of the National Academy of Sciences*, **107**, 21256–62.

Mackay, D.J.C. (2009), *Sustainable Energy – Without the Hot Air*, Cambridge: UIT.

Madison, D. (2007), *Preserving Food Without Freezing or Canning*, Paris: Terre Vivante.

Mair, J. (2011), 'Exploring air travellers' voluntary carbon-offsetting behaviour', *Journal of Sustainable Tourism*, **19**, 215–30.

Makki, A.A., R.A. Stewart, K. Panuwatwanich and C. Beal (2012), 'Revealing the determinants of shower water end use consumption: enabling better targeted urban water conservation strategies', *Journal of Cleaner Production*, DOI: 10.1016/j.jclepro.2011.08.007.

Malpass, A., C. Barnett, N. Clarke and P. Cloke (2007), 'Problematizing choice: responsible consumers and sceptical citizens', in M. Bevin and F. Trentmann (eds), *Governance, Consumers and Citizens*, Basingstoke: Palgrave Macmillan, pp. 231–46.

Manomaivibool, P. (2009), 'Extended producer responsibility in a non-OECD context: the management of waste electrical and electronic equipment in India', *Resources, Conservation and Recycling*, **53**, 136–44.

Mansvelt, J. (2008), 'Geographies of consumption: citizenship, space and practice', *Progress in Human Geography*, **32**, 105–17.

Maréchal, K. (2010), 'Not irrational but habitual: the importance of "behavioural lock-in" in energy consumption', *Ecological Economics*, **69**, 1104–14.

Marias, K. and I.A. Waitz (2009), 'Air transport and the environment', in P. Belobaba, A. Odoni and C. Barnhart (eds), *The Global Airline Industry*, Chichester: Wiley, pp. 405–40.

Marshall, N. and R. Rounds (2011), *Green Burials in Australia and their Planning Challenges*, State of Australian Cities Conference 2011, **29** November-2 December, Melbourne, VIC.

Mason, L., T. Boyle, J. Fyfe, T. Smith and D. Cordell (2011), *National Food Waste Data Assessment: Final Report*, Sydney, NSW: Department of Sustainability, Environment, Water, Population and Communities, Institute for Sustainable Futures, University of Technology.

Mathieu, R., C. Freeman and J. Aryal (2007), 'Mapping private gardens in urban areas using object-oriented techniques and very high-resolution satellite imagery', *Landscape and Urban Planning*, **81**, 179–92.

McDonald, G.W., V.E. Forgie and C. MacGregor (2006), 'Treading lightly: ecofootprints of New Zealand's ageing population', *Ecological Economics*, **56**, 424–39.

McDonald, S., C. Oates, M. Thyne, P. Alevizou and L.A. McMorland (2009), 'Comparing sustainable consumption patterns across product sectors', *International Journal of Consumer Studies*, **33**, 137–45.

McGuirk, P.M. (1997), 'Multiscaled interpretations of urban change: the federal, the state and the local in the Western Area Strategy of Adelaide', *Environment and Planning D*, **15**, 481–98.

McIntosh, A. (2012), 'Ten truths about Australia's rush to mine and the mining workforce', *Australian Geographer*, **43**, 331–7.

McKenny, L. (2012), 'Dilapidated shack at centre of stoush between cemetery and farmers', *Sydney Morning Herald*, 17 April.

McKibben, B. (2012), 'Global warming's terrifying new math', *Rolling Stone*, 2 August, accessed 5 August 2012 at www.rollingstone.com/politics/news-global-warmings-terrifying-new-math-20120719.

McMichael, A.J., R.E. Woodruff and S. Hales (2006), 'Climate change and human health: present and future risks', *Lancet*, **367** (9513), 859–69.

Measham, T. (2006), 'Learning about environments: the significance of primal landscapes', *Environmental Management*, **38**, 426–34.

Mekonnen, M.M. and A.Y. Hoekstra (2010), 'A global and high-resolution assessment of the green, blue and grey water footprint of wheat', *Hydrology and Earth System Sciences*, **14**, 1259–76.

Mila i Canals, L., S.J. Cowell, S. Sim and L. Basson (2007), 'Comparing domestic versus imported apples: a focus on energy use', *Environmental Science and Pollution Research*, **14**, 338–44.

Millennium Ecosystem Assessment (MEA) (2005), *Ecosystems and Human Well-being: Synthesis*, Washington, DC: Island Press.

Miller, D. (1993), 'A theory of Christmas', in D. Miller (ed.), *Unwrapping Christmas*, Oxford: Clarendon Press, pp. 3–37.

Miller, D. (2001), 'Driven societies', in D. Miller (ed.), *Car Cultures*, Oxford: Berg.

Miller, W.I. (1997), *The Anatomy of Disgust*, Cambridge, MA: Harvard University Press.

Ministerial Council on Energy (2008), *National Hot Water Strategic Framework*, Canberra: Department of Resources Energy and Tourism.

Mitchell, T. (2008), 'Rethinking economy', *Geoforum*, **39**, 1116–21.

Mitford, J. (1998), *The American Way of Death Revisited*, New York: Knopf.

Mohanraj, M., S. Jayaraj and C. Muraleedharam (2008), 'Comparative assessment of environment-friendly alternatives to R134a in domestic refrigerators', *Energy Efficiency*, **1**, 189–98.

Molina, M.J. and F.S. Rowland (1974), 'Stratospheric sink for cholor-fluromethanes: chlorine atom-catalysed destruction of ozone', *Nature*, **249**, 810–12.

Monaghan, A. (2009), 'Conceptual niche management of grassroots innovation for sustainability: the case of body disposal practices in the UK', *Technological Forecasting and Social Change*, **76**, 1026–43.

Monbiot, G. (2007), *Heat: How to Stop the Planet Burning*, London: Penguin.

Mooallem, J. (2008), 'The afterlife of cellphones', *New York Times Magazine*, 38–43.

Moody, H.R. (2008), 'Environmentalism as an aging issue', *Public Policy & Aging Report*, **18** (2), 1–7.

Moore, S.A. (2012), 'Garbage matters: concepts in new geographies of waste', *Progress in Human Geography*, DOI: 10.1177/0309132512437077.

Morley, D. (2003), 'What's home got to do with it? Contradictory dynamics in the domestication of technology and the dislocation of domesticity', *European Journal of Cultural Studies*, **6** (4), 435–58.

Morris, J. and J. Bagby (2008), 'Measuring environmental value for natural lawn and garden care practices', *International Journal of Life Cycle Analysis*, **13**, 226–34.

Morrison, M., C.S. McSweeney and A-D.G. Wright (2007), 'The vertebrate animal gut in context, microbiomes, metagenomes and methane', *Microbiology Australia*, 107–10.

Most, E. (2003), *Calling All Cell Phones: Collection, Reuse, and Recycling Programs in the US*, New York: Inform.

Motavalli, J. (2011), *High Voltage: The Fast Track to Plug in the Auto Industry*, Emmaus, PA: Rodale Books.

Moy, C. (2012), 'Rainwater tank households: water savers or water users?', *Geographical Research*, **50**, 204–16.

Murtaugh, P.A. and M.G. Schlax (2009), 'Reproduction and the carbon legacies of individuals', *Global Environmental Change*, **19**, 14–20.

Nansen, B., M. Arnold, M. Gibbs and H. Davis (2011), 'Dwelling with media stuff: latencies and logics of materiality in four Australian homes', *Environment and Planning D*, **29**, 693–715.

Narayanaswamy, V., J. Altham, R. Van Berkel and M. McGregor (2004), *Environmental Life Cycle Assessment (LCA) Case Studies for Western Australian Grain Products*, Curtin University.

Nasar, J.L., J.S. Evans-Cowley and V. Mantero (2007), 'McMansions: the extent and regulation of super-sized houses', *Journal of Urban Design*, **12**, 339–58.

National Greenhouse Gas Inventory (2009), *Australia's National Greenhouse Accounts*, Canberra: Department of Climate Change.

National Recycling Coalition (2007), 'Monitoring the environment', accessed 2 June 2012 at http://nrcrecycles.org/.

Natural Ways (Summer 2007), 'The promise in Promession', accessed 20 December 2011 at www.naturalwayburial.org.uk/page6.htm.

Nelson, A.C. (2010), 'Catching the next wave: older adults and the "New Urbanism"', *Generations*, **33** (4), 37–42.

Newman, P.W.G. and J.R. Kenworthy (1999), *Sustainability and Cities: Overcoming Automobile Dependence*, Washington, DC: Island Press.

Nickles, S. (2002), '"Preserving women": refrigerator design as social process in the 1930s', *Technology and Culture*, **43**, 693–727.

Norman, D.A. (2004), *Emotional Design: Why we Love (or Hate) Everyday Things*, New York: Basic Books.

Nowak, D. and D. Crane (2002), 'Carbon storage and sequestration by urban trees in USA', *Environmental Pollution*, **116**, 381–89.

Nye, D. (1990), *Electrifying America: The Social Meanings of a New Technology*, Cambridge: The MIT Press.

O'Brien, K., R. Olive, Y-C. Hsu, L. Morris, R. Bell and N. Kendall (2009), 'Life cycle assessment: reusable and disposable nappies in Australia', 6th Australian Conference on Life Cycle Assessment, Melbourne (1–14), accessed 24 May 2012 at www.crdc.com.au/uploaded/file/E-Library/Climate%20Change%20July%2009/LCA%20Cotton%20v%20Disposable%20Nappies%20OBrienetal2009.pdf.

OECD (Organisation for Economic Co-operation and Development) (2012), *Transport Outlook 2012: Seamless Transport for Greener Growth*, Paris: OECD.

Office for National Statistics (2000), 'Social Trends', No. 30, accessed 16 August 2012 at www.ons.gov.uk/ons/rel/social-trends-rd/social-trends/no--30--2000-edition/index.html.

OICA (International Organisation of Motor Vehicle Manufacturers) (2012), '2011 Production Statistics', Paris: OICA, accessed 27 September 2011 at http://oica.net/category/production-statistics/.

O'Neill, S.J. and M. Hulme (2009), 'An iconic approach for representing climate change', *Global Environmental Change*, **19**, 402–10.

Ongondo, F.O. and I.D. Williams (2011), 'Greening academia: use and disposal of mobile phones among university students', *Waste Management*, **31**, 1617–34.

Ongondo, F.O., I.D. Williams and T.J. Cherrett (2011). 'How are WEEE doing? A global review of the management of electrical and electronic wastes', *Waste Management*, **31**, 714–30.

Organo, V., L. Head and G. Waitt (2012), 'Who does the work in sustainable households? A time and gender analysis in New South Wales, Australia', *Gender, Place & Culture*, doi: 10.1080/0966369X.2012.716401, available at: www.tandfonline.com/doi/abs/10.1080/0966369X.2012.716401.

Osibanjo, O. and I.C. Nnorom (2008), 'Material flows of mobile phones and accessories in Nigeria: environmental implications and sound end-of-life management options', *Environmental Impact Assessment Review*, **28**, 298–13.

Page, G., B. Ridoutt and B. Bellotti (2012), 'Carbon and water footprint trade-offs in fresh tomato production', *Journal of Cleaner Production*, **32**, 219–26.

Pakula, C. and R. Stamminger (2010), 'Electricity and water consumption for laundry washing by washing machine worldwide', *Energy Efficiency*, **3**, 365–82.

Parker, K., L. Head, L.A. Chisholm and N. Feneley (2008), 'A conceptual model of ecological connectivity in the Shellharbour Local Government Area, New South Wales, Australia', *Landscape and Urban Planning*, **86**, 47–59.

Parliament of South Australia (2008), *Natural Burial Grounds*, Adelaide, SA: Parliament of South Australia, Resources and Development Committee.

Patton, P. (2011), 'How we shower', *Kitchen & Bath Business*, **58** (2), 14–16.

Pauleit, S., R. Ennos and Y. Golding (2005), 'Modeling the environmental impacts of urban land use and land cover change: a study in Merseyside, UK', *Landscape and Urban Planning*, **71**, 295–310.

Pearce, F. (2007), 'Look, no footprint', *New Scientist*, **193** (2594), 38–41.

Pears, K. (2006), *Fashion Re-consumption; Developing a Sustainable Fashion Consumption Practice Influenced by Sustainability and Consumption Theory*, RMIT University, Melbourne, VIC.

Pelletier, N., R. Pirog and R. Rasmussen (2010), 'Comparative life cycle environmental impacts of three beef production strategies in the upper Midwestern United States', *Agricultural Systems*, **103**, 380–89.

Penner, J.E., D.H. Lister, D.J. Griggs, D.J. Dokken and M. McFarland (1999), *Aviation and the Global Atmosphere: A Special Report of IPCC Working Groups I and III in Collaboration with the Scientific Assessment Panel to the Montreal Protocol on Substances that Deplete the Ozone Layer*, Cambridge: Cambridge University Press.

Perez, I. (2012), 'Comfort for the people and liberation for the housewife; gender, consumption and refrigerators in Argentina (1930–60)', *Journal of Consumer Culture*, **12**, 156–74.

Perlin, J. (2002), *From Space to Earth: The Story of Solar Electricity*, Cambridge, MA: Harvard University Press.

Petrini, C.O. (2007), *Slow Food Nation: Why our Food Should be Good, Clean and Fair*, New York: Rixxoli Ex Libris.

Pillemer, K.L. and L. Wagenet (2008), 'Taking action: environmental volunteerism and civic engagement by older people', *Public Policy and Aging Report*, **18**, 23–7.

Pillemer, K., L. Wagenet, D. Goldman, L. Bushway and R. Meador (2010), 'Environmental volunteering in later life: benefits and barriers', *Generations*, **33**, 58–62.

Pillemer, K., N. Wells, L. Wagenet, R. Meador and J. Parise (2011), 'Environmental sustainability in an aging society', *Journal of Aging and Health*, **23**, 433–53.

Pink, S. (2005), 'Dirty laundry. Everyday practice, sensory engagement and the constitution of identity', *Social Anthropology*, **13**, 275–90.

Po, M., J. Kaercher and B. Nancarrow (2004), *Literature Review of Factors Influencing Public Perceptions of Water Reuse*, Perth, WA: CSIRO.

Potter, E. and P. Starr (2006), 'Culture and climate change', *Altitude 7*, accessed 2 September 2012 at www.api-network.com/cgi-bin/altitude21c/fly?page=Issue7&n=8.

Potts, A. (2011), 'Push for "green" cemetery on Coast', *Sun Community Newspapers*, 30 November.

Power, E.R. (2005), 'Human-nature relations in suburban gardens', *Australian Geographer*, **36**, 39–53.

Pritchard, B. and D. Burch (2003), *Agri-Food Globalization in Perspective: International Restructuring in the Processing Tomato Industry*, Aldershot: Ashgate.

Probyn, E. (2010), 'Feeding the world: towards a messy ethics of eating', in T. Lewis and E. Potter, *Ethical Consumption: A Critical Introduction*, New York: Routledge, pp. 103–15.

Pugh, A.J. (2009), *Longing and Belonging: Parents, Children, and Consumer Culture*, Berkeley, CA: University of California Press.

Purtell and Associates (1997), *Improving Consumer Perception of Kangaroo Products*, Canberra: Rural Industries Research and Development Corporation.

Pyne, S. (1995), *World Fire: The Culture of Fire on Earth*, New York: Holt.

Qantas (2012a), 'Carbon pricing', accessed 28 May 2012 at www.qantas.com.au/travel/airlines/carbon-pricing/global/en.

Qantas (2012b), 'Offset my flight now', accessed 17 May 2012 at www.qantas.com.au/travel/airlines/offset-my-flight/global/en.

Rainey, T. (2012), 'Weighing the environmental costs: buy an eReader, or a shelf of books?', *The Conversation*, accessed 26 July 2012 at www.theconversation.edu.au.

Rapier, R. (2012), *Power Plays: Energy Options in the Age Of Peak Oil*, New York: Apress.

Ratnasingam, J. and F. Ioras (2003), 'The sustainability of the Asian wooden furniture industry', *European Journal of Wood and Wood Products*, **61**, 233–7.

Reid, L., P. Sutton and C. Hunter (2010), 'Theorizing the meso level: the household as a crucible of pro-environmental behaviour', *Progress in Human Geography*, **34**, 309–27.

Reinhart, A.K. (1990), 'Impurity/no danger', *History of Religions*, **30**, s1–24.

Ribot, J. (2010), 'Vulnerability does not fall from the sky: toward multiscale, pro-poor climate policy', in R. Mearns and A. Norton (eds), *Social Dimensions of Climate Change*, Washington, DC: World Bank.

Richter, T. (2004), 'Energy efficient laundry processes, GE Global Research', accessed 5 September 2008 at www.osti.gov/energycitations/purl.cover.jsp?purl=/842014-FdebHg/native/.

Riedy, C. and G. Milne (2010), 'Energy use: hot water service', accessed 24 July 2012 at www.yourhome.gov.au/technical/pubs/fs65.pdf

Robbins, P. (2007), *Lawn People. How Grasses, Weeds and Chemicals Make Us Who We Are*, Philadelphia, PA: Temple University Press.

Robbins, P. and J.T. Sharp (2003), 'Producing and consuming chemicals: the moral economy of the American lawn', *Economic Geography*, **74**, 425–71.

Robinson, L.M. (1997), 'Safeguarded by your refrigerator: Mary Engle Pennington's struggle with the National Association of Ice Industries', in S. Stage and V.B. Vincenti (eds), *Rethinking Home Economics: Women and the History of a Profession*, Ithaca, NY: Cornell University Press, pp. 253–70.

Rockefeller, A.A. (1998), 'Civilization and sludge: notes on the history of the management of human excreta', *Capitalism, Nature, Socialism*, **9** (3), 3–18.

Roggeveen, K. (2010), 'Tomato journeys from farm to fruit shop', MSc thesis, University of Wollongong.

Roggeveen, K. (2012), *Tomato Journeys from Farm to Fruit Shop*, Local Environment, DOI: 10.1080/13549839.2012.738653.

Røpke, I. (2009), 'Theories of practice – new inspiration for ecological economic studies on consumption', *Ecological Economics*, **68**, 2490–97.

Røpke, I. (2012), 'The unsustainable directionality of innovation – the example of broadband transition', *Research Policy*, **41** (9), 1631.

Røpke, I., T.H. Christensen and J.O. Jensen (2010), 'Information and communication technologies – a new round of household electrification', *Energy Policy*, **38**, 1764–73.

Rosenbloom, S. (2001), 'Sustainability and automobility among the elderly: an international assessment', *Transportation*, **28**, 375–408.

Rosenquist, L.E.D. (2005), 'A psychosocial analysis of the human-sanitation nexus', *Journal of Environmental Psychology*, **25**, 335–46.

Rosenthal, E. (2008), 'A line in the yard: the battle over the right to dry outside' *New York Times*, 12 April.

Ross, R. (2008), *Clothing: A Global History*, Cambridge: Polity.

Rowland, F.S. (1997), 'Change of atmosphere', *Our Planet: The Magazine of the United Nations Environment Programme*, **9** (2).

Rugg, J. (2006), 'Lawn cemeteries: the emergence of a new landscape of death', *Urban History*, **33**, 213–33.

Rutland, T. and A. Aylett (2008), 'The work of policy: actor networks, governmentality, and local action on climate change in Portland, Oregon', *Environment and Planning D*, **26**, 627–46.

Salagnac, J-L. (2007), 'Lessons from the 2003 heat wave: a French perspective', *Building Research and Information*, **35**, 450–57.

Sampson, S. (2008), 'Category killers and big-box retailing: their historical impact on retailing in the USA', *International Journal of Retail and Distribution Management*, **36**, 17–31.

Sattar, M.A., R. Saidur and H.H. Masjuki (2007), 'Performance investigation of domestic refrigerator using pure hydrocarbons and blends of hydrocarbons as refrigerants', *World Academy of Science, Engineering and Technology*, **29**, 223–8.

Saunders, C., A. Barber and G. Taylor (2006), *Food Miles – Comparative Energy/Emissions Performance of New Zealand's Agriculture Industry*, research report 285, Lincoln, NZ: Agribusiness & Economics Research Unit Lincoln University.

Scarfo, B. (2011), 'Building a more sustainable future for senior living', *Educational Gerontology*, **37**, 466–87.

Scerri, A. (2011), 'Rethinking responsibility? Household sustainability in the stakeholder society', in R. Lane and A. Gorman-Murray (eds), *Material Geographies of Household Sustainability*, Aldershot: Ashgate.

Schipper, L. (2005), 'Automobile fuel; economy and CO^2 emissions in industrialised countries: troubling trends through 2005/6', World Resource Institute Centre for Sustainable Transport, accessed 27 May 2012 at http://pdf.wri.org/automobile-fuel-economy-co2-industrialized-countries.pdf.

Schlunke, A., J. Lewis and S. Fane (2008), *Analysis of Australian Opportunities for More Efficient Toilets*, Institute for Sustainable Futures, Sydney, NSW: University of Technology, Sydney.

Schlünssen, V., P.S. Vinzents, A.B. Mikkelsen and I. Schaumburg (2001), 'Wood dust exposure in the Danish furniture industry using conventional and passive monitors', *Annals of Occupational Hygiene*, **45**, 157–64.

Schmidt, C.W. (2009), 'Carbon offsets: growing pains in a growing market', *Environmental Health Perspectives*, **117**, A62–A68.

Sheller, M. (2004), 'Automotive emotions: feeling the car', *Theory, Culture & Society*, **21**, 221–42.

Shevchenko, O. (2002), 'In the case of fire emergency: consumption, security and the meanings of durables in a transforming society', *Journal of Consumer Culture*, **22**, 147–70.

Shove, E. (2003), *Comfort, Cleanliness and Convenience: The Social Organization of Normality*, Oxford: Berg.

Shove, E. and D. Southerton (2000), 'Defrosting the freezer: from novelty to convenience. A narrative of normalization', *Journal of Material Culture*, **5**, 301–19.

Siegle, L. (2013) 'Are vintage clothes more ethical?' *The Guardian*, 17 March, accessed 17 March at http://m.guardian.co.uk/environment/2013/mar/17/are-vintage-clothes-more-ethical.

Silva, E.B. (2000), 'The cook, the cooker and the gendering of technology', *The Sociological Review*, **48**, 612–28.

Silva, R. and W. Malagutti Filho (in press), 'Geoelectrical mapping of contamination in the cemeteries: the case study in Piracicaba, São Paulo/Brazil', *Environmental Earth Sciences*, 1–13.

Silveira, G.T.R. and S-Y. Chang (2010), 'Cell phone recycling experiences in the United States and potential recycling options in Brazil', *Waste Management*, **30**, 2278–91.

Silverstone, R. and E. Hirsch (1992), *Consuming Technologies: Media and Information in Domestic Spaces*, London: Routledge.

Sivulka, J. (2001), 'Stronger than dirt: a cultural history of advertising personal hygiene in America, 1875 to 1940', *The Journal of American History*, **89**, 1540–41.

Slocum, R. (2004), 'Consumer citizens and the cities for climate protection campaign', *Environment and Planning A*, **36**, 763–82.

Sydney Morning Herald (SMH) (2007), 'Nappy returns', *Sydney Morning Herald*, accessed 17 August 2012 at http://blogs.smh.com.au/science/archives/2007/04/nappy_returns.html.

SMH (2008), 'Cemetery offers carbon neutral funerals', *Sydney Morning Herald*, 11 March.

Smith, R.M., K.J. Gaston, P.H. Warren and K. Thompson (2005), 'Urban domestic gardens (V): relationships between land cover composition, housing and landscape', *Landscape Ecology*, **20**, 235–53.

Sofoulis, Z. (2005), 'Big water, everyday water: a sociotechnical perspective', *Continuum*, **19**, 445–63.

Sofoulis, Z. (2010), *Water Managers' Views on the Social Dimensions of Urban Water*, Centre for Cultural Research, University of Western Sydney.

Southerton, D. (2001), 'Consuming kitchens', *Journal of Consumer Culture*, **1**, 179–203.

Spinney, J (2012), 'Making mobile mothers', accessed 23 August 2012 at http://justinspinneyresearch.co.uk/?page_id=15.

Spinney, J., N. Green, K. Burningham, G. Cooper and D. Uzzell (2012), 'Are we sitting comfortably? Domestic imaginaries, laptop practices, and energy use', *Environment and Planning A*, **44**, 2629–45.

Stanes, E.R. (2008), *Is Green the New Black? Questions of Sustainability within the Fashion Industry*, BSc thesis, University of Wollongong.

Stelter, B. (2011), 'Ownership of TV sets falls in U.S', *New York Times*, 3 May, accessed 4 September 2012 at www.nytimes.com/2011/05/03/business/media/03television.html..

Stephens Jr, M. and T. Unayama (2012), 'The impact of retirement on household consumption in Japan', *Journal of the Japanese and International Economies*, **26**, 62–83.

Stichnothe, H., A. Hospido and A. Azapagic (2008), 'Carbon footprint estimation of food production systems: the importance of considering methane and nitrous oxide', *Aspects of Applied Biology*, **87**, 65–71.

Still, R. (2008), *Reducing the Carbon Footprint of Wastewater Systems*, New South Wales: Department of Commerce.

Strandbakken, P. (2009), 'Sociology fools the technician? Product durability and social constraints to eco-efficiency for refrigerators and freezers', *International Journal of Consumer Studies*, **33**, 146–50.

Strengers, Y. (2008), 'Comfort expectations: the impact of demand-management strategies in Australia', *Building Research & Information*, **36**, 381–91.

Strengers, Y. (2011), 'Negotiating everyday life: the role of energy and water consumption feedback', *Journal of Consumer Culture*, **311**, 319–38.

Suau-Sanchez, P., M. Pallares-Barbera and V. Paül (2011), 'Incorporating annoyance in airport environmental policy: noise, societal response and community participation', *Journal of Transport Geography*, **19**, 275–84.

Sustainable Illawarra (2008), 'History', accessed 14 September 2010 at www.sustainableillawarra.com.au/.

Sustainability Victoria (2011), *2010 Greenlight Report*, Melbourne, VIC: Victorian Government.

The Centre for Natural Burial (2010), 'How much does a green burial cost?', accessed 1 May 2012 at www.naturalburial.coop/2010/12/27/how-much-does-a-green-burial-cost/.

Tilman, D., R. Socolow, J.A. Foley, J. Hill, E. Larsown, L. Lynd, S. Pacala, J. Reilly, T. Searchinger, C. Somerville and R. Williams (2009), 'Beneficial biofuels – the food, energy and environment trilemma', *Science*, **325** (5938), 270–71.

Tomes, N. (1998), *The Gospel of Germs: Men, Women and the Microbe in American Life*, Cambridge, MA and London: Harvard University Press.

Treehugger.com (2012), 'How to go green: babies', accessed 31 August 2012 at www.treehugger.com/htgg/how-to-go-green-babies.html.

Troy, P. (ed.) (2008), *Troubled Waters. Confronting the Water Crisis in Australia's Cities*, ANU E-Press.

Troy, P., D. Holloway and B. Randolph (2005), *Water Use and the Built Environment: Patterns of Water Consumption in Sydney*, Sydney, NSW: University of New South Wales.

Troy, P.N. (1996), *The Perils of Urban Consolidation: A Discussion of Australian Housing and Urban Development Policies*, Sydney, NSW: Federation Press.

Tuan, Y-F. (1977), *Space and Place: The Perspective of Experience*, Minneapolis, MN: University of Minnesota Press.

Tudor, T., G.M. Robinson, M. Riley, S. Guilbert and S.W. Barr (2011), 'Challenges facing the sustainable consumption and waste management agendas: perspectives on UK households', *Local Environment*, **16**, 51–66.

Tukker, A., G. Huppes, J. Guinée, R. Heijungs, A. de Koning, L. van Oers, S. Suh, T. Gerrken, M. Van Hodlerbeke, B. Jansen and P. Neilsen (2006), *Environmental Impacts of Products (EIPRO): Analysis of the Life Cycle Environmental Impacts Related to the Final Consumption of the EU-25*, European Commission, accessed 7 September 2012 at www.jrc.es.

Tynan, C. and S. McKechnie (2009), 'Hedonic meaning creation through Christmas consumption: a review and model', *Journal of Customer Behaviour*, **8**, 237–55.

Uitdenbogerd, D.E., N.M. Brouwer and J.P. Groot-Marcus (1998), *Domestic Energy Saving Potentials for Food and Textiles: An Empirical Study*, Wageningen, Netherlands: Wageningen Agricultural University.

UN (2011), *World Population Prospects*, New York: United Nations.

UN DESA (United Nations Department of Economic and Social Affairs) (2007), *Sustainable Consumption and Production: Promoting Climate-Friendly Household Consumption Patterns*, New York: UN DESA.

UNEP (United Nations Environment Program) (2007), *2006 Report on the Refrigeration, Air Conditioning and Heat Pumps, Technical Options Committee*, Nairobi: UNEP.

UN FAO (United Nations Food and Agriculture Organization) (2011), *Global Food Losses and Food Waste: Extent, Causes and Prevention*, Rome: UN FAO.

United States Energy Independence and Security Act of 2007, Public Law 110–140, H.R. 6, 2007.

Urry, J. (2004), 'The "system" of automobility', *Theory, Culture and Society*, **21**, 25–39.

Vaajasaari, K., M. Kulovaara, A. Joutti, E. Schultz and K. Soljamo (2004), 'Hazardous properties of paint residues from the furniture industry', *Journal of Hazardous Materials*, **106**, 71–9.

van der Horst, D. and S. Vermeylen (2011), 'Spatial scale and social impacts of biofuel production', *Biomass and Bioenergy*, **35**, 2435–43.

Verbeek, P. (2004), *What Things Do: Philosophical Reflections on Technology, Agency and Design*, University Park, PA: Pennsylvania State University Press.

Victorian Government Department of Health (2009), 'Heatwave Plan for Victoria 2009–2010: protecting health and reducing harm from heatwaves',

Melbourne: State Government of Victoria, accessed 14 August 2012 at www.health.vic.gov.au/environment/downloads/heatwave_plan_vic.pdf.

Vigarello, G. (1998), *Concepts of Cleanliness: Changing Attitudes in France Since the Middle Ages*, Cambridge: Cambridge University Press.

Vincent, J. (2006), 'Emotional attachment and mobile phones', *Knowledge, Technology & Policy*, **19**, 39–44.

Vizcarra, A.T., H.V. Lo and P.H. Liao (1994), 'A lifecycle inventory of baby diapers subject to Canadian conditions', *Environmental Toxicology and Chemistry*, **13**, 1707–16.

von Weissäcker, E., A.B. Lovins and L.H. Lovins (1997), *Factor Four. Doubling Wealth, Halving Resource Use*, The New Report to the Club of Rome, London: Earthscan.

Waight, E. (2011), 'The social role of NCT nearly new sales', accessed 24 September 2012 at www.emmawaight.co.uk/website/wp-content/uploads/2011/04/EWaightYear1Poster.pdf.

Waitt, G. and T. Harada (2012), 'Driving, cities and changing climates', *Urban Studies*, DOI: 10.1177/0042098012443858.

Waitt, G., C. Farbotko and B. Criddle (2012a), 'Scalar politics of climate change: regions, emissions and responsibility', *Media International Australia*, **143**, 36–46.

Waitt, G., P. Caputi, C. Gibson, C. Farbotko, L. Head, N. Gill and E. Stanes (2012b), 'Sustainable household capability: which households are doing the work of environmental sustainability?', *Australian Geographer*, **43**, 51–74.

Wajcman, J. (1991), *Feminism Confronts Technology*, Cambridge: Polity Press.

Wallgren, C. (2006), 'Local or global food markets: a comparison of energy use for transport', *Local Environment*, **11**, 233–51.

Walter, T. (2005), 'Three ways to arrange a funeral: mortuary variation in the modern West', *Mortality*, **10**, 173–92.

Walters, J. (2002), 'Save the planet … stay on the ground', *The Observer*, 12 May.

Warburton, J. and M. Gooch (2007), 'Stewardship volunteering by older Australians: the generative response', *Local Environment*, **12**, 43–55.

Ward's Automotive Group (2011), 'Vehicles in operation by country 2010', accessed 12 June 2012 at http://wardsauto.com/datasheet/world-vehicles-operation-vehicle-type-1930-2010.

Warren, A. and C. Gibson (2011), 'Blue-collar creativity: reframing custom-car culture in the imperilled industrial city', *Environment and Planning A*, **43**, 2705–22.

Water UK (2009), 'How the water industry is managing its contribution to climate change', briefing paper, accessed 3 May 2012 at www.water.org.uk.

Watkins, H. (2003), 'Fridge stories: three geographies of the domestic refrigerator', in M. Hard, A. Losch and D. Verdicchio (eds), *Transforming Spaces. The Topological Turn in Technological Studies*, online publication of the international conference held in Darmstadt, Germany, 22–24 March, accessed 14 August 2012 at www.ifs.tu-darmstadt.de/fileadmin/.../space-folder/program-neu.htm.

Watkins, H. (2008), 'Journeys of the domestic refrigerator', PhD thesis, University of British Columbia.

Weinhold, B. (2011), 'Fields and forests in flames: vegetation smoke and human health', *Environmental Health Perspectives*, **119**, 386–93.

Weisbrod, A. and G. Van Hoof (2012), 'LCA-measured environmental improvements in Pampers diapers', *International Journal of Life Cycle Assessment*, **17**, 145–53.

Weiss, W. (ed.) (2003), S*olar Heating Systems for Houses – A Design Handbook for Solar Combisystems*, London: International Energy Agency/James and James Publishers.

Weiss, W. and F. Mauthner (2012), *Solar Heat Worldwide: Markets and Contribution to the Energy Supply 2010*, Paris: Solar Heating and Cooling Programme, International Energy Agency.

Wen-Tien, T. (2005), 'An overview of environmental hazards and exposure and explosive risk of hydroflurocarbon HFCs', *Chemosphere*, **61**, 1539–47.

West, S. (2011), '"Smart" fridges stay cool by talking to each other', accessed 30 June 2012 at www.csiro.au/Portals/Multimedia/CSIROpod/smart-fridges.aspx.

Whitmarsh, L., G. Seyfang and S. O'Neil (2011), 'Public engagement with carbon and climate change: to what extent is the public "carbon capable"?', *Global Environmental Change*, **21**, 56–65.

WHO (World Health Organization) (2012), 'Water supply, sanitation and hygiene development', accessed 27 September 2012 at www.who.int/water_sanitation_health/hygiene/en/.

Wilhelm, W., A. Yankov and P. Magee (2011), 'Mobile phone consumption behavior and the need for sustainability innovations', *Journal of Strategic Innovation and Sustainability*, **7** (2), 20–40.

Wilhelm, W.B. (2012), 'Encouraging sustainable consumption through product lifetime extension: the case of mobile phones', *International Journal of Business and Social Science*, **3** (3), 17–33.

Wilhite, H., H. Nakagami, T. Masuda, Y. Yamaga and H. Haneda (1996), 'A cross-cultural analysis of household energy use behaviour in Japan and Norway', *Energy Policy*, **24**, 795–803.

Wilkerson, J.T., M.Z. Jacobson, A. Malwitz, S. Balasubramanian, R. Wayson, G. Fleming, A.D. Naiman and S.K. Lele (2010), 'Analysis of emission data from global commercial aviation: 2004 and 2006', *Atmospheric Chemistry and Physics*, **10**, 6391–408.

Williams, E.D., R.U Ayres and M. Heller (2002), 'The 1.7 kilogram microchip: energy and material use in the production of semiconductor devices', *Environmental Science & Technology*, **36**, 5504–10.

Williams, K., E. Burton and M. Jenks (2000), 'Achieving sustainable urban form', in K. Williams, E. Burton and M. Jenks (eds), *Achieving Sustainable Urban Form*, London: Spon, pp. 1–6.

Wilson, G.R. and M.J. Edwards (2008), 'Native wildlife on rangelands to minimize methane and produce lower-emission meat: kangaroos versus livestock', *Conversation Letters*, **1**, 119–28.

Wilson, J. and J.L. Grant (2009), 'Calculating ecological footprints at the municipal level: what is a reasonable approach for Canada?', *Local Environment*, **14**, 963–79.

Winward, J., P. Schiellerup and B. Boardman (1998), 'Cool labels', research report 20, Environmental Change Institute, University of Oxford.

World Bank (2005), 'World Development Indicators', accessed 14 July 2012 at http://data.worldbank.org/data-catalog/world-development-indicators.

World Bank (2012), 'World Development Indicators, television ownership data', accessed 12 July 2012 at www.nationmaster.com/graph/med_hou_wit_tel-media-households-with-television#source.

World Toilet Organization (2011), 'World Toilet Organization won Gates Foundation's support of USD270,000', press release 1 November 2011, Singapore, accessed 26 September 2012 at www.worldtoilet.org/WTD/gates_foundation.pdf

WorldWatch (2012), 'China reports 66-percent drop in plastic bag use', accessed July 2012 at www.worldwatch.org/node/6167.

Wright, J., P. Osman and P. Ashworth (2009), *The CSIRO Home Energy Saving Handbook*, Sydney, NSW: Pan Macmillan.

Wright, S.D., M. Caserta, et al. (2003), 'Older adults' attitudes, concerns and support for environmental issues in the "New West"', *International Journal of Aging and Human Development*, **57**, 151–79.

WSSCC (World Supply and Sanitation Collaborative Council) (2012), 'Vital statistics', Geneva: WSSCC, accessed 26 September 2012 at www.wsscc.org/media/vital-statistics.

Wu, B.Y., Y.C. Chan, A. Midendorf, X. Gu and H.W. Zhong (2008), 'Assessment of toxicity potential of metallic elements in discarded electronics: a case study of mobile phones in China', *Journal of Environmental Sciences*, **20**, 1403–8.

Yeandle, S. (1984), *Women's Working Lives: Patterns and Strategies*, London: Tavistock Publications.

Yu, E. and J. Liu (2007), 'Environmental impacts of divorce', *Proceedings of the National Academy of Sciences*, **104**, 20629–34.

Zadok, G. and R. Puustinen (2010), *The Green Switch: Designing for*

Sustainability in Mobile Computing, USENIX Workshop on Sustainable Information Technology, San Jose, CA.

Zasada, I., S. Alves, F.C. Müller, A. Piorr, R. Berges and S. Bell (2010), 'International retirement migration in the Alicante region, Spain: process, spatial pattern and environmental impacts', *Journal of Environmental Planning and Management*, **53**, 125–41.

Zhang, A., S.V. Gudmundsson and T.H. Oum (2010), 'Editorial: air transport, global warming and the environment', *Transportation Research Part D*, **15**, 1–4.

Zong, C-B and K. Liljenquist (2006), 'Washing away your sins: threatened morality and physical cleansing', *Science*, **313**, 1451–2.

Zychowski, J. (2012), 'Impact of cemeteries on groundwater chemistry: a review', *CATENA*, **93**, 29–37.

Index